NOT IN OUR GENES

BIOLOGY, IDEOLOGY, AND HUMAN NATURE

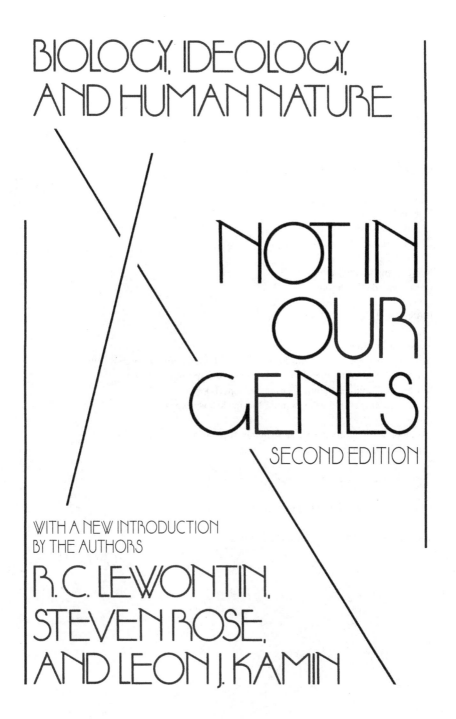

NOT IN OUR GENES

SECOND EDITION

WITH A NEW INTRODUCTION
BY THE AUTHORS

R. C. LEWONTIN, STEVEN ROSE, AND LEON J. KAMIN

HAYMARKET BOOKS
CHICAGO, IL

© 1984 by R. C. Lewontin, Steven Rose, and Leon J. Kamin

Originally published in 1984 by Pantheon Books

This edition published in 2017 by
Haymarket Books
P.O. Box 180165
Chicago, IL 60618
773-583-7884
info@haymarketbooks.org
www.haymarketbooks.org

ISBN: 978-1-60846-727-3

Trade distribution:
In the US through Consortium Book Sales and Distribution, www.cbsd.com
In the UK, Turnaround Publisher Services, www.turnaround-uk.com
In Canada, Publishers Group Canada, www.pgcbooks.ca
All other countries, Publishers Group Worldwide, www.pgw.com

This book was published with the generous support of the Wallace Action
Fund and Lannan Foundation.

Cover design by Josh On.

Library of Congress CIP Data is available.

Entered into digital printing March, 2021.

Men at some time are masters of their fates:
The fault, dear Brutus, is not in our stars,
But in ourselves, that we are underlings. . . .

Julius Caesar I, ii

CONTENTS

PREFACE TO THE 2017 EDITION

We thank Haymarket Books for making *Not in Our Genes* available to a new and probably younger readership. However, this edition comes with a health warning. Our book was first published in 1984 and engages extensively with issues that were particularly relevant to us—a neuroscientist, psychologist, and geneticist—in the United States and United Kingdom in that decade, so they may at first sight seem arcane to a new generation of readers. Nonetheless, we hope that the analyses we develop in *Not in Our Genes* will strike a chord with you.

We were then, and are still, particularly concerned with the ways in which theories and practices in the life sciences and biomedicine articulate with the dominant ideology of late twentieth and now early twenty-first-century capitalism. Looking back over *Not in Our Genes* from a 2016 vantage point, we are struck by how much has changed, yet how much remains the same. Today's fully-fledged globalized neoliberal economy is more than ever wedded to the methodological individualism and reductionism we discussed in *Not in Our Genes*, the critique of which frames every chapter of the book. In the intervening thirty years since the book's publication, the small-scale sciences of genetics and neuroscience have been transformed into today's massive technosciences. Despite—or perhaps because of—these transformations, the determinist arguments that root the massive and growing inequalities in wealth, status, and power that characterize all modern economies in inexorable biological laws retain a strong grasp on popular understanding. Evolutionary, genetic, and neurologically "justified" hierarchies of class, sex, and race remain acute, though sometimes masked in a modern, more ameliorative language.

Not in Our Genes makes a critical analysis of the following claims:

1. The hereditary—that is, genetic—basis of differences in intelligence between individuals and so-called "races"
2. The biologization of patriarchy and gender relations
3. The medicalization of unruly behavior
4. The rise of biological psychiatry
5. The globalizing pretensions of sociobiology

And finally, because critique is insufficient without offering a better alternative theory, we sketch out the lineaments of a different understanding of living processes as radically indeterminate, and how, to quote the final words of the book, it is our biology that makes us free.

It would be a Herculean task to spell out in any detail the ways in which developments in the sciences since the 1980s have affected these critiques, but in brief:

1. The hereditarian arguments around intelligence—IQ—depend, first, on the proposition that intelligence is what IQ tests measure, second, that intelligence is largely genetically determined, and third, that the average difference in IQ scores between "races" are based on genetic differences. *Not in Our Genes* shows that each of these propositions is false. Nothing that has happened in the intervening period alters our conclusion.

i. The critique that we and others made of IQ tests may have helped hasten their demise. Certainly they are out of fashion today in both the United States and United Kingdom.

Nonetheless, that on average poor and working-class children perform worse on IQ tests than middle- and upper-class children do has been used by neoliberal theorists in both the United States and United Kingdom to explain and justify the existence of a permanent underclass, or precariat, lacking what is currently termed "mental capital."

ii. The genetic argument, which in the 1980s was still based largely on the twin studies that we analyze in chapter 4, has been overtaken by the advances in gene sequencing that led, by the turn of the millennium, to the decoding of the human genome. Determinists claimed that the sequencing of the three billion DNA base pairs that constitute the genome would provide the "book of life" in which would be inscribed the fate of any individual. In fact, what the sequencing has shown is that, far from our lives being determined by the 22,000 or so genes within each person's genome, it is how the genes are read and regulated during development (epigenetics) that matters—as we argue in the final chapter of *Not in Our Genes*.

The technical advances of the 1990s that made the Human Genome Project possible have continued ever since, so that a person's entire genome can be sequenced within a week at a price not much above $100. This has opened the way to hunt for specific "intelligence genes." The hunt has been spectacularly unsuccessful; no such major genes have been identified; those that might be involved account for only a small fraction of the heritability. Geneticists have begun to speak of "lost heritability." Others might conclude that the entire genetic paradigm is broken.

iii. With the advances in sequencing came the recognition—as we had insisted in the book, and despite some attempts to use the new genetics to "re-racialize" biology—that traditional broad biology-based racial classifications between, for instance, blacks and whites, were unsustainable. There are, however, differences in gene frequencies between population groups, so that it becomes appropriate to use the rather cumbersome phrase "bio-geographical ancestry."

2. Despite being subject to trenchant and continuing critique by feminist biologists and sociologists, the familiar patriarchal arguments

discussed in chapter 6 continue almost unchanged. The rise of evolution-ary psychology (see point 5 below) has been accompanied by continuing fantasies about the "fixing" of gender relations in the Pleistocene, en-suring, for example, that men preferred to mate with younger women with a body shape indicative of fertility, while women preferred older, dominant, powerful men. The role of hormones in early prenatal devel-opment has been formalized into "Brain Organization Theory," in which a surge of testosterone in about the third trimester is said to masculinize an otherwise female brain, generating what one prominent male author describes as "the essential difference." As feminist critics have spelled out in detail, such biodeterminist arguments are deaf and blind to the mas-sive changes in gender relations that the last half century of technological and social change has wrought.

3. The medicalization of unruly behavior continues apace. The ad-vent of brain imaging techniques such as functional magnetic resonance imaging has led to claims that it is possible to detect brain regions that predispose a person to psychopathy, or even to detect whether they have "terrorist thoughts." Minimal Brain Dysfunction, which chap-ter 7 discusses, is now called Attention Deficit Hyperactivity Disorder (ADHD), and has been joined in the latest version of the U.S. psychi-atrists' diagnostic manual by such other bad behaviors as Opposition Defiant Disorder and Conduct Disorder—labels that could as readily be applied to demonstrators against injustice and state oppression as to cheeky children in class. Meanwhile ADHD, a largely American disease in the 1980s, has spread across the globe, with 5–10 percent of children in countries from Iceland to Australia being diagnosed, and the prescription rates for drugs, predominantly Ritalin, to treat the al-leged symptoms have more than doubled in the last five years. But the suggestion that there is something wrong with the brains of those di-agnosed with ADHD and that Ritalin specifically treats them is clearly false—and indeed Ritalin is widely available via the Internet and used to enhance attention and learning by increasing numbers of "normal" students revising for exams.

4. Just as the medicalization of unruly behavior has expanded, so too have the claims for biological psychiatry, discussed in chapters 7 and 8. The World Health Organization has described an "epidemic" of depression sweeping the globe. Yet for all the intense research in universities and the gigantic pharmaceutical companies (Big Pharma),

and despite the huge advances in genetics, the genes and biochemistry believed to cause schizophrenia and depression remain elusive. "Lost heritability" is as relevant here as it is to IQ. Worse for Big Pharma, there are no new drugs in sight to treat these conditions. There is nothing to improve on the drugs of the 1980s, now long out of patent. So much so that many of the drug companies are withdrawing from the field, falling back to the safer ground of cancer.

5. Other than for humans, sociobiology has ceased to be so controversially all-embracing, becoming one approach among many in the ethological literature to explain social behavior and mating strategies, though it is still driven by the adaptationist perspective we criticize in *Not in Our Genes*. Meanwhile human sociobiology has changed its name, rebranding itself as evolutionary psychology, its adherents now arguing that human nature—whatever we may mean by that—was fixed in the Pleistocene, and that there has not been enough evolutionary time since then for it to change. Thus we have "Stone Age minds in Internet Age bodies." The one redeeming feature of this shift has been that unlike sociobiology, evolutionary psychology stresses human universals, and therefore has no space for racist typology—although it still leaves plenty of room for sexual stereotyping (see point 2 above). Apart from bones, cave paintings, and the few artifacts found in burial mounds, we have no real idea of the social or psychological lives of our Pleistocene ancestors; thus, there are no limits, other than their ideological predispositions, to the evolutionary psychologists' speculations. But they are wrong about the rate of evolutionary change; rapid changes in human physiology and behavior within a few thousand years have now been well documented, and it is more appropriate to speak of the co-evolution of human biology and culture than of an impossible evolutionary stasis.

In conclusion, to expand the closing words of the book, more than ever today it is true that humans "make their own history, but they do not make it as they please; they do not make it under self-selected circumstances but under circumstances existing already"—circumstances biological, historical, technological, cultural, and social.

—Steven Rose, Leon J. Kamin, and Richard C. Lewontin
August 2016

PREFACE AND ACKNOWLEDGMENTS

The authors of *Not in Our Genes* are respectively an evolutionary geneticist, a neurobiologist, and a psychologist. Over the past decade and a half we have watched with concern the rising tide of biological determinist writing, with its increasingly grandiose claims to be able to locate the causes of the inequalities of status, wealth, and power between classes, genders, and races in Western society in a reductionist theory of human nature. Each of us has been engaged for much of this time in research, writing, speaking, teaching, and public political activity in opposition to the oppressive forms in which determinist ideology manifests itself. We share a commitment to the prospect of the creation of a more socially just—a socialist—society. And we recognize that a critical science is an integral part of the struggle to create that society, just as we also believe that the social function of much of today's science is to hinder the creation of that society by acting to preserve the interests of the dominant class, gender, and race. This belief—in the possibility of a critical and liberatory science—is why we have each in our separate ways and to varying degrees been involved in the development of what has become known over the 1970s and 1980s, in the United States and Britain, as the radical science movement.

The need was, we felt, for a systematic exploration of the scientific and social roots of biological determinism, an analysis of its present-day social functions, and an exposure of its scientific pretensions. More than that, though, it was also necessary to offer a perspective on what biology and psychology can offer as an alternative, a liberatory, view of the "nature of human nature." Hence, *Not in Our Genes*.

The book has been several years in the making. This has been in part a consequence of working at several thousand miles' separation yet wanting to produce an integrated, coherent account rather than a series of separate chapters. As well, the long gestation period has enabled us to develop our own ideas from the initial critical task to the more synthetic statement of the final chapter. This process was crucially aided by the continued testing of our ideas in practice, in debate, in polemic, and in campaign over the period of the production of the book. For one of us (Steven Rose), participation in the remarkable experience of the Dialectics of Biology conference held in Bressanone, Italy, in April 1980 was extremely helpful to this endeavor. Much of the writing was done over a period spent by one of us (again Steven Rose) as a visiting scholar at the Museum of Comparative Zoology at Harvard and in intensive sessions among the authors in Vermont and Maine and their English equivalent, Wharfedale.

Each of us owes intellectual and emotional debts to lovers, comrades, colleagues, teachers, and students. Inevitably these debts are only partially acknowledged by mentioning names here or by citations in the references at the end of the book. And inevitably, neither we nor those we mention may always be aware of the extent and the ways in which their ideas and the discussions we have had together with them have helped shape our thinking.

But we would like particularly to mention: members of the Dialectics of Biology Group and the Campaign Against Racism, I.Q., and the Class Society, Martin Barker, Mike Cooley, Stephen Gould, Agnes Heller, Ruth Hubbard, Phillip Kitcher, Richard Levins, Mary Jane Lewontin, Eli Messinger, Diane Paul, Benjamin Rose, Hilary Rose,

Michel Schiff, Peter Sedgwick, and Ethel Tobach. Needless to say, they are responsible only for any of the virtues and none of the vices of what follows.

Endless draft manuscripts were typed by Jane Bidgood and Beverley Simon at the Open University, Becky Jones at Harvard, and Elaine Bucsik at Princeton. Laurie Melton, of the Open University library, identified innumerable obscure references.

Finally, our thanks to our publishers—Pantheon, Penguin, and Mondadori—for their forbearance.

NOT IN OUR GENES

CHAPTER
ONE

THE NEW RIGHT
AND THE OLD
DETERMINISM

The New Right and the

Old Biological Determinism.

The start of the decade of the 1980s was symbolized, in Britain and the United States, by the coming to power of new conservative governments; and the conservatism of Margaret Thatcher and Ronald Reagan marks in many ways a decisive break in the political consensus of liberal conservatism that has characterized governments in both countries for the previous twenty years or more. It represents the expression of a newly coherent and explicit conservative ideology* often described as the New Right.[1]

*We should make it clear that we use the term *ideology* here and throughout this book with a precise meaning. Ideologies are the ruling ideas of a particular society at a

New Right ideology has developed in Europe and North America in response to the gathering social and economic crises of the past decade. Abroad, in Africa, Asia, and Latin America, there have been struggles against nationalist forces determined to throw off the yoke of political and economic exploitation and colonialism. At home, there has been increasing unemployment, relative economic decline, and the rise of new and turbulent social movements. During the sixties and early seventies, Europe and North America experienced an upsurge of new movements, some of which were quite revolutionary: struggles of shop-floor workers against meritocratic ruling elites, blacks against white racism, women against patriarchy, students against educational authoritarianism, welfare clients against the welfare bureaucrats. The New Right criticizes the liberal response to these challenges of the previous decades, the steady increase in state intervention, and the growth of large institutions, resulting in individuals losing control over their own lives, and hence an erosion of the traditional values of self-reliance which the New Right regards as characterizing the Victorian laissez-faire economy. This movement has been strengthened, in the later seventies and eighties, by the fact that liberalism has fallen into a self-confessed disarray, leaving the ideological battlefield relatively open to the New Right.

The response of the liberal consensus to challenges to its institutions has always been the same: an increase in interventive programs of social amelioration; of projects in education, housing, and inner-city renewal. By contrast, the New Right diagnoses the liberal medicine as merely adding to the ills by progressively eroding the "natural" values that had characterized an earlier phase of capitalist industrial society. In the words of the conservative theoretician Robert Nisbet, it is a reaction against the present-day "erosion of traditional authority in kinship, locality, culture, language, school, and other elements of the social fabric."[3]

particular time. They are ideas that express the "naturalness" of any existing social order and help maintain it:

The ideas of the ruling class are in every epoch the ruling ideas; i.e. the class which is the ruling material force of society is at the same time its ruling intellectual force. The class which has the means of material production at its disposal has control at the same time over the means of mental production, so that thereby, generally speaking, the ideas of those who lack the means of mental production are subject to it. The ruling ideas are nothing more than the ideal expression of the dominant material relationships.[2]

But New Right ideology goes further than mere conservatism and makes a decisive break with the concept of an organic society whose members have reciprocal responsibilities. Underlying its *cri de coeur* about the growth in state power and the decline in authority—underlying even the monetarism of Milton Friedman—is a philosophical tradition of individualism, with its emphasis on the priority of the individual over the collective. That priority is seen as having both a moral aspect, in which the rights of individuals have absolute priority over the rights of the collectivity—as, for example, the right to destroy forests by clear-cutting in order to maximize immediate profit—and an ontological aspect, where the collectivity is nothing more than the sum of the individuals that make it up. And the roots of this methodological individualism lie in a view of human nature which it is the main purpose of this book to challenge.

Philosophically this view of human nature is very old; it goes back to the emergence of bourgeois society in the seventeenth century and to Hobbes's view of human existence as a *bellum omnium contra omnes*, a war of all against all, leading to a state of human relations manifesting competitiveness, mutual fear, and the desire for glory. For Hobbes, it followed that the purpose of social organization was merely to regulate these inevitable features of the human condition.[4] And Hobbes's view of the human condition derived from his understanding of human biology; it was biological inevitability that made humans what they were. Such a belief encapsulates the twin philosophical stances with which this book is concerned, and to which, in the pages that follow, we will return again and again.

The first is *reductionism*—the name given to a set of general methods and modes of explanation both of the world of physical objects and of human societies. Broadly, reductionists try to explain the properties of complex wholes—molecules, say, or societies—in terms of the units of which those molecules or societies are composed. They would argue, for example, that the properties of a protein molecule could be uniquely determined and predicted in terms of the properties of the electrons, protons, etc., of which its atoms are composed. And they would also argue that the properties of a human society are similarly no more than the sums of the individual behaviors and tendencies of the individual humans of which that society is composed. Societies are "aggressive" because the individuals who compose them are "aggressive," for instance. In formal language, reductionism is the claim that the compositional units of a whole are ontologically prior to the whole that the units comprise. That is, the units and their properties exist *before* the whole, and there is a chain of causation that runs from the units to the whole.[5]

The second stance is related to the first; indeed, it is in some senses a special case of reductionism. It is that of *biological determinism.* Biological determinists ask, in essence, Why are individuals as they are? Why do they do what they do? And they answer that human lives and actions are inevitable consequences of the biochemical properties of the cells that make up the individual; and these characteristics are in turn uniquely determined by the constituents of the genes possessed by each individual. Ultimately, all human behavior—hence all human society—is governed by a chain of determinants that runs from the gene to the individual to the sum of the behaviors of all individuals. The determinists would have it, then, that human nature is fixed by our genes. The good society is either one in accord with a human nature to whose fundamental characteristics of inequality and competitiveness the ideology claims privileged access, or else it is an unattainable utopia because human nature is in unbreakable contradiction with an arbitrary notion of the good derived without reference to the facts of physical nature. The causes of social phenomena are thus located in the biology of the individual actors in a social scene, as when we are informed that the cause of the youth riots in many British cities in 1981 must be sought in "a poverty of aspiration and of expectation created by family, school, environment, and genetic inheritance."[6]

What is more, biology, or "genetic inheritance," is always invoked as an expression of inevitability: What is biological is given by nature and proved by science. There can be no argument with biology, for it is unchangeable, a position neatly exemplified in a television interview given by British Minister for Social Services Patrick Jenkin in 1980 on working mothers:

Quite frankly, I don't think mothers have the same right to work as fathers. If the Lord had intended us to have equal rights to go to work, he wouldn't have created men and women. These are biological facts, young children do depend on their mothers.

The use of the double legitimation of science and god is a bizarre but not uncommon feature of New Right ideology: the claim to a hotline to the deepest sources of authority about human nature.

The reductionist and biological determinist propositions that we shall examine and criticize in the pages of this book are:

- Social phenomena are the sums of the behaviors of *individuals*.
- These behaviors can be treated as objects, that is, *reified* into properties located in the brain of particular individuals.
- The reified properties can be measured on some sort of scale so that individuals can be ranked according to the amounts they possess.
- Population norms for the properties can be established: Deviations from the norm for any individual are *ab*normalities that may reflect medical problems for which the individual must be treated.
- The reified and medicalized properties are *caused* by events in the brains of individuals—events that can be given anatomical localizations and are associated with changed quantities of particular biochemical substances.
- The changed concentrations of these biochemicals may be partitioned between genetic and environmental causes; hence the "degree of inheritance" or *heritability* of differences may be measured.
- Treatment for abnormal amounts of the reified properties may be either to eliminate undesirable genes (eugenics, genetic engineering, etc.); or to find specific drugs ("magic bullets") to rectify the biochemical abnormalities or to excise or stimulate particular brain regions so as to eliminate the site of the undesirable behavior. Some lip service may be paid to supplementary environmental intervention, but the primary prescription is "biologized."

Working scientists may believe, or conduct experiments, based on one or more of these propositions without feeling themselves to be full-fledged determinists in the sense that we use the term; nonetheless, adherence to this general analytical approach characterizes determinist methodology.

Biological determinism *(biologism)* has been a powerful mode of explaining the observed inequalities of status, wealth, and power in contemporary industrial capitalist societies, and of defining human "universals" of behavior as natural characteristics of these societies. As such, it has been gratefully seized upon as a political legitimator by the New Right, which finds its social nostrums so neatly mirrored in nature; for if these inequalities are biologically determined, they are therefore inevitable and immutable. What is more, attempts to remedy them by social means, as in the prescriptions of liberals, reformists, and revolutionaries, "go against nature." Racism, Britain's National Front tells us, is a product of our "selfish genes."[7] Nor are such political dicta confined to the ideologues: Time and again, despite their professed belief that their science is "above mere human politics" (to quote Oxford sociobiologist Richard Dawkins),[8] biological determinists deliver themselves of social and political judgments. One example must suffice for now: Dawkins himself, in his book *The Selfish Gene*, which is supposed to be a work on the genetic basis of evolution and which is used as a textbook in American university courses on the evolution of behavior, criticizes the "unnatural" welfare state where

we have abolished the family as a unit of economic self-sufficiency and substituted the state. But the privilege of guaranteed support for children should not be abused. . . . Individual humans who have more children than they are capable of raising are probably too ignorant in most cases to be accused of conscious malevolent exploitation. Powerful institutions and leaders who deliberately encourage them to do so seem to me less free from suspicion.[9]

The point is not merely that biological determinists are often somewhat naive political and social philosophers. One of the issues with which we must come to grips is that, despite its frequent claim to be neutral and objective, science is not and cannot be above "mere" human politics. The complex interaction between the evolution of scientific theory and the evolution of social order means that very often the ways in which scientific research asks its questions of the human and natural worlds it proposes to explain are deeply colored by social, cultural, and political biases.[10]

Our book has a two fold task: we are concerned first with an explanation of the origins and social functions of biological determinism in general—the task of the next two chapters—and second with a systematic examination and exposure of the emptiness of its claims vis-à-vis the nature and limits of human society with respect to equality, class, race, sex, and "mental disorder." We shall illustrate this through a study of specific themes: IQ theory, the assumed basis of differences in "ability" between sexes and races, the medicalization of political protest, and, finally, the overall conceptual strategy of evolutionary and adaptationist explanation offered by sociobiology in its modern forms. Above all, this means an examination of the claims of biological determinism regarding the "nature of human nature."

In examining these claims and in exposing the pseudoscientific, ideological, and often just simply methodologically inadequate findings of biological determinism, it is important for us, and for our readers, to be clear about the position we ourselves take.

Critics of biological determinism have frequently drawn attention to the ideological role played by apparently scientific conclusions about the human condition that seem to flow from biological determinism. That, despite their pretensions, biological determinists are engaged in making political and moral statements about human society, and that their writings are seized upon as ideological legitimators, says nothing, in itself, about the scientific merits of their claims.[11] Critics of biological determinism are often accused of merely disliking its political conclusions. We have no hesitation in agreeing that we do dislike these conclusions; we believe that it is possible to create a better society than the one we live in at present; that inequalities of wealth, power, and status are not "natural" but socially imposed obstructions to the building of a society in which the creative potential of all its citizens is employed for the benefit of all.

We view the links between values and knowledge as an integral part of doing science in this society at all, whereas determinists tend to deny that such links exist—or claim that if they do exist they are exceptional pathologies to be eliminated. To us such an assertion of the separation of fact from value, of practice from theory, "science" from "society" is itself part of the fragmentation of knowledge that reductionist thinking sustains and which has been part of the mythology of the last century of "scientific advance" (see Chapters 3 and 4). However, the least of our tasks here is that of criticizing the social implications of biological determinism, as if the broad claims of biological determinism could be upheld. Rather, our major goal is to show that the world is not to be understood as biological determinism would have it be, and that, as a way of explaining the world, biological determinism is fundamentally flawed.

Note that we say "the world," for another misconception is that the criticism of biological determinism applies only to its conclusions about human societies, while what it says about nonhuman animals is more or less valid. Such a view is often expressed—for instance about E. O. Wilson's book *Sociobiology: The New Synthesis*,[12] which we discuss at length in Chapter 9. Its liberal critics claim that the problem with *Sociobiology* lies only in the first and last chapters, where the author discusses human sociobiology; what's in between is true. Not so, in our view: what biological determinism has to say about human society is more wrong than what it says about other aspects of biology because its simplifications and misstatements are the more gross. But this is not because it has developed a theory applicable only to nonhuman animals; the method and theory are fundamentally flawed whether applied to the United States or Britain today, or to a population of savanna-dwelling baboons or Siamese fighting fish.

There is no mystical and unbridgeable gulf between the forces that shape human society and those that shape the societies of other organisms; biology is indeed relevant to the human condition, although the form and extent of its relevance is far less obvious than the pretensions of biological determinism imply. The antithesis often presented as an opposition to biological determinism is that biology stops at birth, and from then on culture supervenes. This antithesis is a type of cultural determinism we would reject, for the cultural determinists identify narrow (and exclusive) causal chains in society which are in their own way reductionist as well. Humanity cannot be cut adrift from its own biology, but neither is it enchained by it.

Indeed, one may see in some of the appeal of biological determinist and New Right writing a reassertion of the "obvious" against the very denial of biology that has characterized some of the utopian writings and hopes of the revolutionary movements of the past decade. The post-1968 New Left in Britain and the United States has shown a tendency to see human nature as almost infinitely plastic, to deny biology and acknowledge only social construction. The helplessness of childhood, the existential pain of madness, the frailties of old age were all transmuted to mere labels reflecting disparities in power.[13] But this denial of biology is so contrary to actual lived experience that it has rendered people the more ideologically vulnerable to the "common-sense" appeal of reemerging biological determinism. Indeed, we argue in Chapter 3 that such cultural determinism can be as oppressive in obfuscating real knowledge about the complexity of the world we live in as is biological determinism. We do not offer in this book a blueprint or a catalogue of certainties; our task, as we see it, is to point the way toward an integrated understanding of the relationship between the biological and the social.

We describe such an understanding as dialectical, in contrast to reductionist. Reductionist explanation attempts to derive the properties of wholes from intrinsic properties of parts, properties that exist apart from and before the parts are assembled into complex structures. It is characteristic of reductionism that it assigns relative weights to different partial causes and attempts to assess the importance of each cause by holding all others constant while varying a single factor. Dialectical explanations, on the contrary, do not abstract properties of parts in isolation from their associations in wholes but see the properties of parts as arising out of their associations. That is, according to the dialectical view, the properties of parts and wholes codetermine each other. The properties of individual human beings do not exist in isolation but arise as a consequence of social life, yet the nature of that social life is a consequence of our being human and not, say, plants. It follows, then, that dialectical explanation contrasts with cultural or dualistic modes of explanation that separate the world into different types of phenomena—culture and biology, mind and body—which are to be explained in quite different and nonoverlapping ways.

Dialectical explanations attempt to provide a coherent, unitary, but nonreductionist account of the material universe. For dialectics the universe is unitary but always in change; the phenomena we can see at any instant are parts of processes, processes with histories and futures whose paths are not uniquely determined by their constituent units. Wholes are composed of units whose properties may be described, but the interaction of these units in the construction of the wholes generates complexities that result in products qualitatively different from the component parts. Think, for example, of the baking of a cake: the taste of the product is the result of a complex interaction of components—such as butter, sugar, and flour—exposed for various periods to elevated temperatures; it is not dissociable into such-or-such a percent of flour, such-or-such of butter, etc., although each and every component (and their development over time at a raised temperature) has its contribution to make to the final product. In a world in which such complex developmental interactions are always occurring, history becomes of paramount importance. Where and how an organism is now is not merely dependent upon its composition at this time but upon a past that imposes contingencies on the present and future interaction of its components.

Such a world view abolishes the antitheses of reductionism and dualism; of nature/nurture or of heredity/environment; of a world in stasis whose components interact in fixed and limited ways, indeed in which change is possible only along fixed and previously definable pathways. In the chapters that follow, the explication of this position will appear in the course of the development of our opposition to biological determinism—in our analysis, for instance, of the relationship of genotype and phenotype (in Chapter 5), and of mind and brain.

Let us take just one example here, that of the relationship of the organism to its environment. Biological determinism sees organisms, human or nonhuman, as adapted by evolutionary processes to their environment, that is, fitted by the processes of genetic reshuffling, mutation, and natural selection to maximize their reproductive success in the environment in which they are born and develop; further, it sees the undoubted plasticity of organisms—especially humans—as they develop as a series of modifications imposed upon an essentially passive, recipient object by the buffeting of "the environment" to which it is exposed and to which it must adapt or perish. Against this we counterpose a view not of organism and environment insulated from one another or unidirectionally affected, but of a constant and active interpenetration of the organism with its environment. Organisms do not merely receive a given environment but actively seek alternatives or change what they find.

Put a drop of sugar solution onto a dish containing bacteria and they will actively move toward the sugar till they arrive at the site of optimum concentration, thus changing a low-sugar for a high-sugar environment. They will then actively work on the sugar molecules, changing them into other constituents, some of which they absorb, others of which they put out into the environment, thereby modifying it, often in such a way that it becomes, for example, more acid. When this happens, the bacteria move away from the highly acid region to regions of lower acidity. Here, in miniature, we see the case of an organism "choosing" a preferred environment, actively working on it and so changing it, and then "choosing" an alternative.

Or consider a bird building a nest. Straw is not part of the bird's environment unless it actively seeks it out so as to construct its nest; in doing so it changes its environment, and indeed the environment of other organisms as well. The "environment" itself is under constant modification by the activity of all the organisms within it. And to any organism, all others form part of its "environment"—predators, prey, and those that merely change the landscape it resides in.[14]

Even for nonhumans, then, the interaction of organism and environment is far from the simplistic models offered by biological determinism. And this is much more the case for our own species. All organisms bequeath to their successors when they die a slightly changed environment; humans above all are constantly and profoundly making over their environment in such a way that each generation is presented with quite novel sets of problems to explain and choices to make; we make our own history, though in circumstances not of our own choosing.

It is precisely because of this that there are such profound difficulties with the concept of "human nature." To the biological determinists the old credo "You can't change human nature" is the alpha and omega of the explanation of the human condition. We are not concerned to deny that there *is* a "human nature," which is simultaneously biologically and socially constructed, though we find it an extraordinarily elusive concept to pin down; in our discussion on sociobiology in Chapter 9 we analyze the best list of human "universals" that protagonists of sociobiology have been able to present.

Of course there *are* human universals that are in no sense trivial: humans are bipedal; they have hands that seem to be unique among animals in their capacity for sensitive manipulation and construction of objects; they are capable of speech. The fact that human adults are almost all greater than one meter and less than two meters in height has a profound effect on how they perceive and interact with their environment. If humans were the size of ants, we would have an entirely different set of relations with the objects that constitute our world; similarly, if we had eyes that were sensitive, like those of some insects, to ultraviolet wavelengths, or if, like some fishes, we had organs sensitive to electrical fields, the range of our interactions with each other and with other organisms would doubtless be very different. If we had wings, like birds, we would construct a very different world.

In this sense, the environments that human organisms seek and those they create are in accord with their nature. But just what does this mean? The human chromosomes may not contain the genes that, in the development of the phenotype, are associated with ultraviolet vision, or sensitivity to electrical fields, or wings. Indeed, in the last case there are structural reasons quite independent of genetic ones why organisms of the weight of humans cannot develop wings large or powerful enough to enable them to fly. And indeed, for a considerable proportion of human history it has gone against human nature to be able to do any of these things. However, as is apparent to all of us, in our present society we can do all of these things: see in the ultraviolet; detect electrical fields; fly by machine, wind, or even pedal power. It is, clearly, "in" human nature to so modify our environment that all these activities come well within our range (and hence within the range of our genotype).

Even where the acts we perform on our environment appear to be biologically equivalent, they are not necessarily socially equivalent. Hunger is hunger (the anthropologist Lévi-Strauss has made this given the basis of a complex human structural typology); yet hunger satisfied by eating raw meat with hands and fingers is quite different from that satisfied by eating cooked meat with a knife and fork. All humans are born, most procreate, all die; yet the social meanings invested in any of these acts vary profoundly from culture to culture and from context to context within a culture.

This is why about the only sensible thing to say about human nature is that it is "in" that nature to construct its own history. The consequence of the construction of that history is that one generation's limits to the nature of human nature become irrelevant to the next. Take the concept of intelligence. To an earlier generation, the capacity to perform complex long multiplication or division was laboriously acquired by children fortunate enough to go to school. Many never achieved it; they grew up lacking, for whatever reason, the ability to perform the calculations. Today, with no more than minimal training, such calculating power and considerably more are at the disposal of any five-year-old who can manipulate the buttons on a calculator. The products of one human generation's intelligence and creativity have been placed at the disposal of a subsequent generation, and the horizons of human achievement have been thereby extended. The intelligence of a schoolchild today, in any reasonable understanding of the term, is quite different from and in many ways much greater than that of his or her Victorian counterpart, or that of a feudal lord or of a Greek slaveowner. Its measure is itself historically contingent.

Because it is in human nature so to construct our own history, and because the construction of our history is made as much with ideas and words as with artifacts, the advocacy of biological determinist ideas, and the argument against them, are themselves part of that history. Alfred Binet, the founder of IQ testing, once protested against "the brutal pessimism" that regards a child's IQ score as a fixed measure of his or her ability, rightly seeing that to regard the child as thus fixed was to help ensure that he or she remained so. Biological determinist ideas are part of the attempt to preserve the inequalities of our society and to shape human nature in their own image. The exposure of the fallacies and political content of those ideas is part of the struggle to eliminate those inequalities and to transform our society. In that struggle we transform our own nature.

CHAPTER TWO

THE POLITICS OF BIOLOGICAL DETERMINISM

When Oliver Twist first meets young Jack Dawkins, the "Artful Dodger," on the road to London, a remarkable contrast in body and spirit is established. The Dodger was a "snub-nosed, flat-browed, common-faced boy enough . . . with rather bow legs and little sharp ugly eyes." And as might be expected from such a specimen, his English was not of the nicest: " 'I've got to be in London tonight,' " he tells Oliver, " 'and I know a 'spectable old genelman as lives there, wot'll give you lodgings for nothink. . . .' " We can hardly expect more from a ten-year-old boy of the streets with no family, no education, and no companions except the lowest criminals of the London *lumpenproletariat*. Or can we? Oliver's manner is genteel and his speech perfect. " 'I am very hungry and tired,' " says Oliver, "the tears standing in his eyes as he spoke. 'I have walked a long way. I have been walking these seven days.' " He was a "pale, thin child," but there was a "good sturdy spirit in Oliver's breast." Yet Oliver was raised from birth in that most degrading of nineteenth-century British institutions, the parish workhouse, with no mother, no education. During the first nine years of his life, he, together with "twenty or thirty other juvenile offenders against the poor-laws, rolled about the floor all day, without the inconvenience of too much food or too much clothing." Where, amid the oakum pickings, did Oliver garner that sensitivity of soul and perfection of English grammar that was the complement to his delicate physique? The answer, which is the solution to the central mystery that motivates the novel, is that Oliver's blood was upper middle class even though his nourishment was gruel. Oliver's father was the scion of a well-off and socially ambitious family; his mother was the daughter of a naval officer. Oliver's life is a constant affirmation of the power of nature over nurture. It is a nineteenth-century version of the modern adoption study showing that children's temperamental and cognitive traits resemble those of their biological parents even though they are placed at birth in an orphanage. Blood will tell, it seems.

Dickens' explanation of the contrast between Oliver and the Artful Dodger is one form of the general ideology of biological determinism as it has developed in the last 150 years into an all-encompassing theory that goes well beyond the assertion that an individual's moral and intellectual qualities are inherited. It is, in fact, an attempt at a total system of explanation of human social existence, based on the two principles that human social phenomena are the direct consequences of the behaviors of individuals, and that individual behaviors are the direct consequences of inborn physical characteristics. Biological determinism is, then, a reductionist explanation of human life in which the arrows of causality run from genes to humans and from humans to humanity. But it is more than mere explanation: It is politics. For if human social organization, including the inequalities of status, wealth, and power, are a direct consequence of our biologies, then, except for some gigantic program of genetic engineering, no practice can make a significant alteration of social structure or of the position of individuals or groups within it. What we are is natural and therefore fixed. We may struggle, pass laws, even make revolutions, but we do so in vain. The natural differences between individuals and among groups played out against the background of biological universals of human behavior will, in the end, defeat our uninformed efforts to reconstitute society. We may not live in the best of all *conceivable* worlds, but we live in the best of all *possible* worlds.

As we have said, for the past fifteen years in America and Britain, and more recently elsewhere in Western Europe, biological determinist theories have become an important element in political and social struggles. The beginning of the most recent wave of biologistic explanation of social phenomena was Arthur Jensen's article in the *Harvard Educational Review* in 1969 arguing that most of the difference between blacks and whites in their performance on IQ tests was genetic.[1] The conclusion for social action was that no program of education could equalize the social status of blacks and whites, and that blacks ought better to be educated for the more mechanical tasks to which their genes predisposed them. Quite soon the claim of the genetic inferiority of blacks was extended to the working class in general and given wide popular currency by another professor of psychology, Richard Herrnstein, of Harvard.[2] The determinist thesis was immediately incorporated into discussions on public policy. Daniel P. Moynihan, the advocate in the American government of "benign neglect" of the poor, felt the winds of Jensenism blowing through Washington. The Nixon administration, anxious to find justifications for severe cuts in expenditures on welfare and education, found the genetic argument particularly useful.

In Britain, promoted by yet a third academic psychologist, Hans Eysenck, the claim of biological differences in IQ between races has become an integral part of the campaign against Asian and black immigration.[3] The purported intellectual inferiority of immigrants simultaneously explains their high rate of unemployment and their demands upon the public welfare apparatus, and justifies restricting their further immigration. Moreover, it legitimizes the racism of the fascist National Front, which argues in its propaganda that modern biology has proven the genetic inferiority of Asians, Africans, and Jews.

A second direction of biological determinist argument that has direct political consequences is the explanation of the domination of women by men. In the last ten years claims for basic biological differences between the sexes in temperament, cognitive ability, and "natural" social role have played an important part in the struggle against the political demands of the women's movement. The successful campaign to prevent ratification of the Equal Rights Amendment to the Constitution of the United States made extensive use of the claims of sociobiologists for the immutability of male social supremacy. At the peak of the struggle for the Equal Rights Amendment, the most widely read newspapers and magazines in America gave prominence to the views of academic biologists like E. O. Wilson of Harvard, who assured his readers that "even in the most free and most egalitarian of future societies men are likely to continue to play a disproportionate role in political life, business, and science."[4]

While biological determinism claims immutability for those characteristics of human behavior that are universal or for differences in social status between larger groups, it also prescribes biological cures for sporadic deviance. If genes cause behavior, then bad genes cause bad behavior, and a cure for social pathology lies in fixing defective genes. Thus, a third political direction of biological determinism has been as a mode of explanation of "social deviance" and, in particular, violence. The uprisings of blacks in American cities, the organized and individual revolts of prisoners, the crimes of personal violence that are said to be increasing in frequency all contribute to a consciousness of violence that demands a defense in the form of "law and order" and an explanation that cites a causal route specific enough to justify the defense. Biological determinism locates the defect in the brains of individuals. Deviant behavior is seen as the result of a deviant organ of behavior; the appropriate treatment is by pill or knife. Large numbers of prisoners have been "cured" of their social deviance by drugs and by the conditioning methods of animal behavior psychology. Moreover, a general application of psychosurgery and psychopharmaceutics is the recommended response to a general outbreak of violence. Thus, psychosurgeons Mark and Ervin argue in their book *Violence and the Brain*[5] that as only *some* blacks in American ghettos participated in the numerous uprisings of the 1960s and 1970s, the social conditions, to which *all* were exposed, cannot be the cause of their violence. The violent cases were those with diseased brains and should be so treated.

But overt violence is not the only manifestation of the diseased brain for which determinists offer a biological explanation and treatment. Children from whom the schools elicit only boredom or impatient fidgeting or distraction are "hyperactive" or suffer "minimal brain dysfunction." Again a disordered brain is seen as the cause of an unacceptable interaction of individuals and social organizations. The political consequence is that, since the social institution is never questioned, no alteration in it is therefore contemplated; individuals are to be altered to fit the institutions or else sequestered to suffer in isolation the consequences of their defective biology.

Most recently, an extension has been made from disordered brain to defective body. It is now clear that certain work hazards—for example noxious chemicals, high noise levels, and electromagnetic radiation— are responsible for a great deal of chronic illness including permanent respiratory disorders, nervous disorders, and cancer. While the first obvious response to this knowledge is to alter the conditions of work in favor of the worker, it is now being suggested seriously that workers be screened for susceptibility to pollutants before they are hired. Those who are "excessively" susceptible would be denied employment.[6]

All of these recent political manifestations of biological determinism have in common that they are directly opposed to the political and social demands of those without power. The postwar period in Britain and America, especially in the last twenty-five years, has been marked by increasing militancy on the part of groups that had previously made few pressing demands. This militancy was in part a consequence of economic and social changes produced by the Second World War. In Britain, Asians and Africans of the new Commonwealth countries were encouraged to immigrate to relieve the severe labor shortage. In the United States large numbers of blacks and women had been incorporated into the industrial work force and armed forces. But the postwar economic boom was short-lived, and, by the late 1950s in Britain and the early 1960s in America, economic difficulties began. Asians and Africans who had previously been perceived as foreign subject races by Britons were now visible immigrants demanding jobs and social services from a shrinking economy. Black militancy in America grew even as the economy cooled. In both countries there was a strong sense that an embattled majority was under constant siege from an unstable minority. In the United States, black militancy radicalized unexpected groups—for example, prisoners—and challenged, threateningly, fundamental assumptions about the inherent good—or primacy—of the existing order. Black radical intellectuals like Malcolm X changed the interpretation of crime and imprisonment from individual social pathology into a form of political struggle. If "all property is theft," then theft is a just form of redistribution of property, a view reechoed in the summer riots of 1981 in Britain. Independent labor militancy was instigated by blacks in industrial firms, in Britain and the United States, a labor militancy that was hostile both to the employers and to the traditional trade union movement that conspired to make blacks the last hired and first fired.

The possibility of profound change moved into nontraditional areas, with new centers of agitation. Mass militancy of women began in the 1960s to exert serious pressures on employers, trade unions, and the state. The work-in movement in failing British light industries, the organization of service workers in hospitals, and the creation of the welfare rights organizations in the United States were largely the work of women—and, in the latter cases, black women.[7] The welfare rights movement transformed support payments to women and dependent children from a dole to be received silently into a right to be demanded loudly.

The 1960s were marked, in general, by an extraordinary breakdown of a previously accepted consensus and an increase in social struggle. Those arrested increasingly demanded rights against police and guards who were seen as oppressive and violent. Students challenged the legitimacy of their universities and schools, and masses of young Americans denied the right and power of the state to conscript them. Consumer and ecological organizations challenged the right of private capital to organize production without reference to public welfare and have demanded state regulation of the process of production.

The declining relative prosperity of Britain beginning in the 1950s and of America in the 1960s made it more and more difficult to accede to the economic pressure of immigrants, blacks, and women. Irrespective of prosperity, neither private capital nor the state that is largely reflective of its interests can afford to yield substantial power and survive. In the end, the owners of capital must control the process of production; the state must control the police and courts; the schools and universities must control the curricula and students.

The growth of biological determinist thought and argument in the early 1970s was precisely a response to the militant demands that increasingly could not be met. It was an attempt to deflect the force of their pressure by denying their legitimacy. The demand by blacks for equal economic reward and social status, it is claimed, is illegitimate because blacks are biologically less capable of dealing with the high abstractions that bring high rewards. The demand of women for equality is unwarranted because male domination has been built into our genes by generations of evolution. The demand of parents for a restructuring of schools to educate their illiterate children cannot be met because their children have dysfunctional brains. The violence of blacks against the property of landlords and merchants is not the outcome of the powerlessness of the propertyless but the consequence of brain lesions. For each militancy, there is an appropriately tailored biological explanation that deprives it of its legitimacy. Biological determinism is a powerful and flexible form of "blaming the victim."[8] As such, we must expect it to become more prominent and diversified as the consciousness of victimization grows while the possibility of accommodating to demands shrinks.

On the other hand, it does not recede entirely when militancy cools. The ten years just preceding the publication of this book have seen some reduction in social unrest in Europe and North America over previous decades. While the renaissance of interest in IQ, genetics, and race, the invention of a sociobiological theory of human nature, and the explicit linkage of social violence with brain disorders all belong to the earlier, more turbulent, era, the production of determinist theory has continued up to the present. In part, this reflects the fact that the production of ideas has a life of its own, given impetus by social events but unfolding through a process that is given by the social organization of intellectual life. Having proposed that blacks are genetically inferior to whites in cognitive skills, Jensen and Eysenck must further develop this theme in response to criticism and in pursuit of the justification that their public personae and their careers demand. Once E. O. Wilson launched his sociobiological theory of human nature, the publication of a series of works by other authors seeking to exploit the evident appeal of the theory became inevitable.

In part, however, the continued production and popularity of biological determinist works, irrespective of the immediate intensity of social struggle, is a consequence of a long-standing contradiction in our society that is in constant need of resolution. The manifest inequalities of status, wealth, and power that characterize society are in patent contradiction to the myth of liberty, equality, and fraternity by which the social order is justified. Biological determinism speaks directly to this inequality and justifies it as natural or fair or both. Any understanding of the roots of biological determinism must then go back to the roots of bourgeois society.

Literary and Scientific Fictions

Despite its claims to a new scientificity, biological determinism has a long history. Since the nineteenth century it has held both a literary and a scientific, although no less fictional, currency. Emile Zola's Rougon-Macquart novels were "experimental novels" meant to show the inevitable consequences of certain scientific facts. In particular, the "facts" were that an individual's life was a result of the unfolding of a hereditary predisposition, and, although environment might temporarily modify its ontogenetic course, in the end heredity triumphed. Gervaise, the laundress in *L'Assommoir* had, by her own exertions, pulled herself out of poverty and was the owner of a thriving business, but one day as she sat, arms immersed in dirty laundry, "her face bent over the bundles, a lassitude seized her . . . as if she were drunk on this human stench, vaguely smiling, eyes glazed. It seemed as if her first sloth arose here, in the asphyxia of the dirty linen polluting the air around her." She had returned to type, to the affinity for degradation and filth that passed into her blood from her drunken layabout father, Antoine Macquart. Her daughter was Nana, who already at the age of five engaged in lewd and vicious games, and who grew up to be a prostitute. Nana's father, Coupeau, when admitted to the hospital for alcoholism, is asked first by the examining physician, "Did your father drink?" The Rougons and Macquarts are two halves of a family descended from a woman whose first, lawful, husband was the solid peasant Rougon, while her second, her lover, was the violent, unstable criminal Macquart. From these two unions there then arose the excitable, ambitious, successful Rougon line and the depraved, alcoholic, criminal Macquarts, among whom are numbered Gervaise and Nana. As Zola says in his preface to the novels, "Heredity has its laws, just as does gravitation."[9]

At first sight there seems to be an inconsistency. The theme of the self-made man who is able, by his own exertions, to break the social bonds that held his ancestors is the one we have come to associate with the bourgeois revolutions of the eighteenth century and the liberal reforms of the nineteenth. Surely if those revolutions meant anything it was the rejection of the principle that merit was hereditary and its replacement by the idea of beginning anew in each generation a freely competitive pursuit of happiness. Zola was a socialist, a republican, and a fierce opponent of inherited privilege. He was notoriously anticlerical, and his famous defense of Dreyfus had as its target the aristocratic class of monarchist officers. There can be no question, in the case of Zola, of literary inconsistency. His commitment to hereditary determination of "feelings, desires, passions, all human manifestations" was an integral part of a world view that was characteristic of an antiaristocratic, anticlerical, radical bourgeois of the Third Republic. It was, as we will argue in detail in Chapter 4, both an attempt to reconcile the facts of an unequal and hierarchical society with the ideology of freedom and equality, and the logical outcome of the reductionist mode of thinking about the world that has been characteristic of science since the bourgeois revolution.

Zola's Rougon-Macquart novels were based upon the scientific claims of Lombroso and Broca that inherited physical characteristics were determinative of mental and moral traits. The Rougon-Macquarts in their turn seem to be the literary prototype for the good and evil Kallikaks,[10] an invented family whose supposed history of inherited vices and virtues graced college psychology texts throughout a large portion of the present century. At present the impression is given by modern biological determinists that the simple objective facts of modern science force us to the conclusion that biology is destiny. The same claim was made in the nineteenth century for Lombroso's criminal anthropology. While no one now would give serious consideration to Lombroso's idea that one can tell a murderer by the shape of his or her head,[11] it is now said that one can do so by the shape of his or her chromosomes. There is an unbroken line of science from the criminal anthropology of 1876 to the criminal cytogenetics of 1975,[12] yet the evidence and argument of determinist claims remain as weak now as they were a hundred years ago. The "scientific" branch of the progressivist hereditarian view flowed together with social Darwinism in the obsessive fear of the deterioration of the "national stock" from the excessive breeding of the working classes. Francis Galton and his protégé Karl Pearson in Britain in the late nineteenth century and early twentieth century began the eugenics movement, which campaigned energetically for selective breeding through the first three decades of the present century. In conformity with their beliefs that differences in abilities could be quantified and partitioned, they developed a host of multifactorial statistical techniques that are the cornerstones of the area of genetical research known since Pearson's day as biometry.[13]

It is important to understand that at times in the history of eugenics in Britain and the United States progressivist movements espoused biological determinism. The early twentieth-century Fabian socialists in Britain, including such figures as G. B. Shaw and the Webbs, were also social imperialists who believed in white superiority and the manifest destiny of the British "race" to encompass the globe.

Since the British were sure that biology was on their side and that the Anglo-Saxons showed genetic superiority over all other "races," the main concern outside socialist circles was with the biology of social class. In the hands of Cyril Burt, a pupil of Pearson, the instruments of quantification of human differences by way of the IQ test and Burt's conviction that IQ differences were largely hereditary (to say nothing of his propensity to invent the "evidence" to support such claims; see Chapter 5) became powerful weapons for restructuring the educational system in specific class interests, as, for example, in the creation of the "eleven-plus" examination that guaranteed the segregation of working-class children into inferior schools from which there was virtually no access to universities.

In the United States the concern of the eugenicists remained overwhelmingly with race differences. It is true that social Darwinism itself was even more extensively used as a legitimator of unrestrained capitalism than in Britain. The ideologue of social Darwinism, Herbert Spencer, was far more influential in the United States, and no one perhaps has captured the spirit of social Darwinism more clearly than John D. Rockefeller, who said at a business dinner, "The growth of a large business is merely a survival of the fittest. . . . This is not an evil tendency in business. It is merely the working out of a law of nature."[14] Nonetheless, in expanding America with its large new immigrant population, it was the racial dimension that was crucial to the social Darwinist and eugenic ideologues, and they included a generation of psychologists who were profoundly to shape the direction that the behavioral sciences were to take from the 1920s onward, with their reductionist convictions that the crucial questions for psychology to answer concerned the origins of individual and group differences in performance.

In 1924 the Congress of the United States passed an immigration restriction act that weighted future immigration in the United States heavily against Eastern and Southern Europeans. Testimony before Congress by leaders of the American mental testing movement to the effect that Slavs, Jews, Italians, and others were mentally dull and that their dullness was racial, or at least constitutional, gave scientific legitimacy to the law that was constructed.[15] Ten years later the same argument was the basis for the German racial and eugenic laws that began with the sterilization of the mentally and morally undesirable and ended in Auschwitz. The claims of biological determinists and eugenicists to scientific respectability were severely damaged in the gas chambers of the "Final Solution." Yet forty years after Burt and thirty years after the start of the 1939–45 war, Arthur Jensen resurrected the hereditarian arguments, uniting the British concern with class and the American obsession with race. At present, the National Front in Britain and the Nouvelle Droite[16] in France argue that racism and anti-Semitism are natural and cannot be eliminated, citing as their authority E. O. Wilson of Harvard, who claims that territoriality, tribalism, and xenophobia are indeed part of the human genetic constitution, having been built into it by millions of years of evolution.

Biological determinists have argued historically that whether or not there may be a pernicious political result of their doctrines is irrelevant to the objective issues about nature. Louis Agassiz, a professor of zoology at Harvard and America's most distinguished zoologist of the nineteenth century, wrote that "we have a right to consider the questions growing out of man's physical relations as merely scientific questions, and to investigate them without reference to politics or religion."[17] The sentiment was echoed in 1975 by yet another Harvard professor and biological determinist, Bernard Davis, who assures us that "neither religious nor political fervor can command the laws of nature."[18] True, but political fervor can apparently command what Harvard professors *say* about the laws of nature, since the eminent zoologist Agassiz claimed that "the brain of the negro is that of the imperfect brain of a seven months infant in the womb of the white"[19] and that the skull sutures of black babies closed earlier than those of whites, so it was impossible to teach black children very much because their brains could not grow beyond the limited capacity of their skulls.

Certainly the repugnant political consequences that have repeatedly flowed from determinist arguments are not criteria by which to judge their objective truth. We cannot derive "ought" from "is" or "is" from "ought," nor will we try (although biological determinists repeatedly do so, as for example E. O. Wilson's demand for a "genetically accurate and hence completely fair code of ethics"[20]). The errors of the biological determinists' explanation of the world can be explicated and understood without reference to the political uses to which these errors have been put. A large part of what follows in this book is an explication of these errors. What cannot be understood without reference to political events, however, is how these errors arise, why they come to characterize both the popular and scientific consciousness in a particular era, and why we should care about them in the first place. We cannot understand Louis Agassiz's extraordinary intellectual dishonesty in claiming as facts things which were not known to be facts, until we read in parts of his memoirs, suppressed until recently, of his total repugnance for and antipathy to blacks, dating from his first arrival in America. He "knew" they were little better than apes when he first laid eyes on them.

Biological determinists try to have it both ways. To legitimize their theories they deny any connection to political events, giving the impression that the theories are the outcome of internal developments within a science that is insulated from social relations. They then become political actors, writing for newspapers and popular magazines, testifying before legislatures, appearing as celebrities on television to explicate the political and social consequences that must flow from their objective science. They change their personae from the scientific to the political and back again as the occasion demands, taking their legitimacy from science and their relevance from politics. They understand that, although there is no logical necessity connecting the truth of determinism to its political role, their own legitimacy as scientific authorities is dependent upon their appearance as politically disinterested parties. In this sense, biological determinists are victims of the very myth of the separation of science from social relations that they and their academic predecessors have perpetuated.

The Role of Scientists

An important feature of biological determinism as a political ideology is its claim to be scientific. Unlike the political philosophy of, say, Plato, whose claims about the nature of society derive from common-sense logical application of certain a prioris, biological determinism claims to be the consequence of modern scientific investigation of the material nature of the human species. It is in the spirit of the *Encyclopedia* of Diderot and d'Alembert, for whom scientific rationality was the basis for all knowledge. As we said in Chapter 1, its closest antecedent in political philosophy is Hobbes, not only because of his adoption of the competitive model for human nature, but also because he was so firmly a mechanistic materialist who derived his political philosophy from assertions about the atomistic notion of individuals in society. Even literary manifestations of determinism, like Zola's, draw their inspiration from the findings of science, although Zola was unusual in his explicit reference to anthropology and his deliberate creation of "experimental" novels.

The characteristic of science, as opposed to prerevolutionary natural philosophy, is that it is an activity of a special group of self-validating experts: scientists. The word "scientist" itself did not come into the language until 1840. The appeal to the "scientific" for legitimacy and to scientists as the ultimate authorities is quintessentially modern. The objectification of social relations which is embodied in science is translated into the objectivity, disinterestedness, and lack of passion of scientists (except their "passion for the truth"). Since science is now the source of legitimacy for ideology, so scientists become the generators of the concrete form in which it enters public consciousness. Since, in the twentieth century, research science, as opposed to development, is carried out chiefly in universities and their allied institutions, it is universities that have become the chief institutions for the creation of biological determinism. But, of course, universities are not merely research institutions. They train the staff who will teach in polytechnic colleges, institutions of higher education without research programs, and community colleges. They train some proportion of secondary and primary school teachers directly, or else the staff for teacher training institutions. And they train, directly, the upper echelons of the middle class. Newspapers, magazines, and televisions all look to the universities as sources of expert knowledge and of "informed opinion." Thus, universities serve as creators, propagators, and legitimators of the ideology of biological determinism. If biological determinism is a weapon in the struggle between classes, then the universities are weapons factories, and their teaching and research faculties are the engineers, designers, and production workers. Over and over again in this book we will analyze the work and quote the conclusions from among the most eminent, successful, and respected of our scientists and professors. Some of what they say will seem ludicrous and some of it deeply shocking. It is important to understand that biological determinism, even in its most gross and vicious forms, is not the product of a fringe of crackpots and vulgar popularizers, but of some of the core members of the university and scientific community. The Nobel Prize laureate Konrad Lorenz, in a scientific paper on animal behavior in 1940 in Germany during the Nazi extermination campaign said:

The selection of toughness, heroism, social utility . . . must be accomplished by some human institutions if mankind in default of selective factors, is not to be ruined by domestication induced degeneracy. The racial idea as the basis of the state has already accomplished much in this respect.[21]

He was only applying the view of the founder of eugenics, Sir Francis Galton, who sixty years before wondered that "there exists a sentiment, for the most part quite unreasonable, against the gradual extinction of an inferior race."[22] What for Galton was a gradual process became rather more rapid in the hands of Lorenz's efficient friends. As we shall see, Galton and Lorenz are not atypical.

Biological Determinism and "Bad Science"

Some critics of biological determinism try to dismiss it as merely bad science. And if the manipulation of data to agree with already held convictions, the deliberate suppression of known facts, the use of simple illogical propositions, and the creation of fraudulent data from nonexistent experiments are universally agreed to be outside the pale of admitted science, then there has been a large amount of "bad science" in support of biological determinism. However, the problem is a great deal more complicated.

"Science" is sometimes taken to mean the body of scientists and the set of social institutions in which they participate, the journals, the books, the laboratories, the professional societies and academies through which individuals and their work are given currency and legitimacy. At other times "science" stands for the set of methods that are used by scientists as means for investigating the relations among things in the world, and the canons of evidence that are accepted as giving credibility to the conclusions of scientists. Yet a third meaning given to "science" is the body of facts, laws, theories, and relationships concerning real phenomena that the social institutions of "science," using the methods of "science," claim to be true.

It is extremely important for us to distinguish what the social institutions of science, using the methods of science, *say* about the world of phenomena from the actual world of phenomena itself. Just because those social institutions, using these methods, have so often said true things about the world, we are in danger of forgetting that sometimes the claims of those who speak in the name of "science" are rubbish.

Why, then, are they given such serious attention? It is because, in contemporary Western society, science as an institution has come to be accorded the authority that once went to the Church. When "science" speaks—or rather when its spokesmen (and they generally are men) speak in the name of science—let no dog bark. "Science" is the ultimate legitimator of bourgeois ideology. To oppose "science," to prefer values to facts, is to transgress not merely against a human law but against a law of nature.

Let us be clear as to what it is we are maintaining about science and its claims: We are *not* arguing that to state the political philosophy or social position of the exponents of a particular scientific claim is enough to evaporate or invalidate that claim. Explaining its origins does not explain away the claim itself. (This is what philosophers call the "genetic fallacy.") We *are* arguing that there are two distinct questions to be asked of any description or explanation offered of the events, phenomena, and processes that occur in the world around us. The first is about the internal logic and asks: Is the description accurate and the explanation true? That is, do they correspond to the reality of the phenomena, events, and processes in the real world?* It is this type of question about the internal logic of science that most Western philosophers of science believe, or claim to believe, science to be all about. The model of scientific advance that most scientists are taught, and which is largely based on the writing of philosophers like Karl Popper and his acolytes, sees science as progressing in this abstract way, by a continuous sequence of theory-making and testing, conjectures and refutations. In the more up-to-date, Kuhnian version of the model, these conjectures and refutations of "normal" science are occasionally convulsed by periods of "revolutionary" science in which the entire framework ("paradigm") within which the conjectures and refutations are framed is shaken, like a kaleidoscope which relocates the same pieces of data into quite new patterns, even though the whole process of theory-making is believed to occur autonomously without reference to the social framework in which science is done.[23]

But the second question to be asked of descriptions or explanations is about the social matrix in which science is embedded—and it is a question of equal importance. The insight into the theories of scientific growth hinted at in the nineteenth century by Marx and Engels, developed by a generation of Marxist scholars in the 1930s, and now reflected, refracted, and plagiarized by a host of sociologists, is that scientific growth does not proceed in a vacuum. The questions asked by scientists, the types of explanation accepted as appropriate, the paradigms framed, and the criteria for weighing evidence are all historically relative. They do not proceed from some abstract contemplation of the natural world as if scientists were programmable computers who neither made love, ate, defecated, had enemies, nor expressed political views.[24]

*To ask this question is to enter a philosophical minefield which surrounds the concept of truth and which we will avoid by offering an essentially operational definition that is appropriate for assessing statements of truth in science, at least. In this definition, a true statement about an event, phenomenon, or process in the real material world must be (a) capable of independent verification by different observers; (b) internally self-consistent; (c) consistent with other statements about related events, phenomena, or processes; and (d) capable of generating verifiable predictions, or hypotheses, about what will happen to the event, phenomenon, or process if it is operated upon in certain ways—if we act upon it.

It is from this perspective that one can see that the internalist, positivist tradition of the autonomy of scientific knowledge is itself part of the general objectification of social relations that accompanied the transition from feudal to modern capitalist societies. That objectification results in a person's status and role in society being determined by his or her relation to objects, while the way in which individuals confront each other is seen as the accidental product of these relations. In particular, scientists are seen as individuals confronting an external and objective nature, wrestling with nature to extract its secrets, rather than as people with particular relations to each other, to the state, to their patrons, and to the owners of wealth and production. Thus, scientists are defined as those who do science, rather than science being defined as what scientists do. But scientists have done more than simply participate in the general objectification of society. They have raised that objectification to the status of an absolute good called "scientific objectivity." Just as the objectification of society in general unleashed the immense productive forces of capitalism, so scientific objectivity in particular was a progressive step toward gaining real knowledge about the world. Such objectivity, as we all recognize, has been responsible for an immense increase in the power to manipulate the world for human purposes. But the emphasis on objectivity has masked the true social relations of scientists with each other and with the rest of society. By denying these relations, scientists make themselves vulnerable to a loss of credibility and legitimacy when the mask slips and the social reality is revealed.

Thus, at any historical moment, what pass as acceptable scientific explanations have both social determinants and social functions. The progress of science is the product of a continuous tension between the internal logic of a method of acquiring knowledge that professes correspondence with and truth about the real material world, and the external logic of these social determinants and functions. Those conservative philosophers who deny the latter, and some more currently fashionable sociologists who wish to dissolve away the former entirely, alike fail to understand the power and role of this tension, which forms the essential dynamic of a science whose ultimate tests are always twofold: tests of truth and of social function.

It follows that to call the science produced by some of the most prestigious, best-funded, most honored, and most status-laden scientists in a field "bad science" demands that we erect some ideal of scientific work whose qualities are derived not from the practice of science but from an abstract philosophy. A major emphasis of one area of Western psychological research for more than fifty years has been the creation of tests to measure a cognitive ability that is thought to be an intrinsic property of each individual. A large fraction of human genetics research has been the study of the heritability of temperamental and mental traits, including their chromosomal basis. The newest form of biological determinism, sociobiology, has been legitimated as a separate field of research with the creation of dozens of new academic positions for "sociobiologists" and the publication of brand-new journals devoted to its subject matter. Science that is broadly funded, that is subject to the scrutiny of journal reviewers and academic selection committees, and whose practitioners are awarded knighthoods, fellowships of the Royal Society, and National Medals of Science is, in one sense of its portmanteau of meanings, simply "science."

If, among mathematicians writing in established journals, it were argued that $1 + 1 = 3$, then that would be what they meant by "mathematics," not "bad mathematics," although, of course, no sensible person would use such a rule for building a house. The problem for an understanding of biological determinism is then not simply to sort "bad science" from "good science," although something of that is involved when questions of fraud arise, but rather to ask how the methodology, conceptualization, and rhetoric of a large part of a "normal" science can be in such poor correspondence with the real world of objective relations that it is intended to reveal. Why do biological determinists use the concepts of nature and nurture as separate causes when developmental genetics long ago showed them to be inseparable? Why do they use statistical methodologies in ways that have been shown by their inventors to be invalid? Why do they carry out experiments without any controls? Why, in their logic, do they take causes for effects, correlations for causations, constants for variables?

Still, it may be argued, if biological determinism is not "bad science" it is, at least, "backward science," "uncritical science," or "soft science" as opposed to the "hard science" of physics and molecular biology. It is not the best that science has to offer, and it may be hoped by continual criticism and education that its practitioners will be brought around to a more rigorous stance. Again, there is some truth in the argument. Just as some of the claims of biological determinism have been invalidated by the revelation of the "bad science," the deliberate frauds and manipulations, so also much of the rest can be and is being disabled by a more rigorous approach to experiment and the logic of inference.

As we will show in some detail, the canons of proof or even of reasonable doubt as they have come to be accepted in human behavioral genetics, sociobiology, and human biopsychology are distinctly less rigorous than those operating in closely allied fields. Minuscule samples, uncontrolled experiments, exquisite analyses of heterogeneous data, and unsupported speculations in place of measurements are all common features of biological determinist literature. For example, the study of the heritability of human intelligence is a special branch of biometrical genetics. Yet paper after paper published in the leading journals of human and behavioral genetics, edited by and refereed by leading human geneticists, commit the most elementary errors in experimental design and analysis, which would never be tolerated in, say, the *Agronomy Journal* or *Animal Science*. To write about human beings gives one a license not extended to the study of corn. *Quod licet Jovi non licet bovi!*

But our criticism of biological determinism is at a more fundamental level: The "bad science" and the "soft science" that characterize the study of human social behavior are the ineluctable consequences of what determinists regard as the questions to be asked. Determinists are committed to the view that individuals are ontologically prior to society and that the characteristics of individuals are a consequence of their biology. The evidence of that prior commitment is, as we will show, overwhelming. The open question for determinists, to the extent that there has been one, is the degree of determination of various traits, and how these traits might be manipulated by means of or in spite of their biology. For a very large number of biological determinists even the question of degree has not been at issue, and their concern appears to have been simply to generate evidence to support their determinist convictions. In either case, "soft" or even "bad" science becomes a means to the end. By a process of "the willing suspension of disbelief," unspoken agreement on an appropriate level of criticality occurs among the interested parties, and a corpus of scientific knowledge is created, validated, and legitimated by its creators. It is not enough, then, to criticize the product. We must first look for the source of the ideology that the product reflects, an ideology which, as we shall show in the next chapter, became a central aspect of bourgeois society as it emerged from European feudalism in the seventeenth century, and has dominated it ever since.

CHAPTER THREE

BOURGEOIS IDEOLOGY AND THE ORIGIN OF DETERMINISM

It is hard to realize today the extent to which primary social relations in early feudal European society, lay between person and person rather than between persons and things. The relationships between lord and vassal, seigneur and serf, entailed mutual obligations that did not depend on an equitable exchange but were absolute on each party separately. Relations to material things—to wealth, land, tools, products, and the range of social activity of each individual, including work obligations, freedom of movement, and freedom to buy and sell—were an indissoluble whole determined for each person by the single fact of status relation. Serfs were tied to the land, but lords could not eject them because their connection to the land flowed from their social status. Once renewable at the death of either lord or vassal, fiefdoms gradually became hereditary and the arrangement they dictated inescapable.

Underlying this social system and legitimizing it was the ideology of grace and, later, of divine right. People held their position in the social hierarchy as the result of the conferral or denial of God's grace; kings claimed their absolute right to rule on the same basis. As grace was inherited through blood, the conferral of grace on the founder of a line was a sufficient *primum mobilum*, guaranteeing grace to biological heirs (although only if legitimate) and insuring stable social and economic relations within and between generations. Changes in position in the social hierarchy, like that of the noble Norman house of Bellême, which arose from a crossbowman of Louis d'Outre-Mer, were explained as being the result of conferrals or withdrawals of grace. Charles I was king of England, *Dei gratia*, but, as Cromwell wryly observed, grace had been removed from him, as evidenced by his severed head.

This static world of social relations legitimated by God reflected, and was reflected by, the dominant view of the natural world as itself static. Unlike the more modern view of an essentially progressive and changing world, the feudal universe was conceived as being held in an ever revolving daily and seasonal dance, with the sun, moon, and stars rotating like bright lights fixed to a series of crystal spheres at the center of which was our earth, on which humans themselves were the central part of God's creation. Nature and humanity existed in order to serve God and God's representatives on earth, the lords temporal and spiritual.

In such a world, social and natural change alike were to be discouraged. Just as the heavenly spheres were fixed, so was the social order. People knew their place, were born and lived in it; it was natural and, like nature itself, ever changing on the mundane, quotidian level and yet basically immutable in the larger scheme. In this precapitalist world, not yet dominated by the metaphor of the machine (in which all phenomena are reduced to their component cogs and pulleys, linked in linear chains of cause and effect), it was possible to be much more tolerant of apparently contradictory or overlapping explanations. The causes of events did not have to be mutually consistent. Sickness could be a natural phenomenon in its own right or a visitation from the Lord. Objects were not individual, atomistic, and separate but fluid and varied, and could be transformed one into another. People could become wolves, lead transmute into gold, fair foul and foul fair. It was possible to believe, at one and the same time, both that living forms had each been created separately according to Biblical myth and had existed unchanged since those Edenic days, *and* that individuals were mutable. Myths abounded of hybrid beasts, half horse, half human; and of women who gave birth to monsters as a result of impressions fixed by some event during pregnancy.

Humanity's relationship to nature was not one of domination—because the appropriate machinery of domination did not exist—but rather of coexistence, which demanded respect for and integration with the natural world within which human lives were embedded. This nature was static in the long run and capricious in the short, and any understanding of it, then, could not be based in the end on constant manipulation and transformation, the active techniques of scientific experimentation, but had to be expressed as passive appreciation. Explanations therefore were couched in terms of appeals to the authority of ancient writings, Biblical or Greek, and not on empirical data.

The Rise of Bourgeois Society

It is clear that feudal society was quite unsuited to a growing mercantile, manufacturing, and eventually capitalist system. First, social and economic life had to become disarticulated so that each individual could play many different roles, confronting others sometimes as buyer, sometimes as seller; sometimes as producer, sometimes as consumer; and sometimes as owner, sometimes as user. The particular role played came to depend upon a momentary relation to objects of production and exchange, not upon lifelong social relations.

Second, individuals had to become "free," but only in particular senses. Ties to specific places or persons had to be eliminated, freeing workers to leave land and lord in order to become manufacturing laborers and to move about in commerce. Reciprocally, landowners had to be free to alienate the land, ejecting inefficient and unproductive systems of production. The enclosure acts that began in Britain as early as the thirteenth century and reached a peak in the late seventeenth and eighteenth centuries were designed to concentrate large tracts of land into intensively cultivated and grazed holdings. A consequence of the dispossession of tenants was the creation of a large mobile army of prospective wage laborers for a growing industry. Freedom also had to come in terms of the ownership of one's own body, what Macpherson calls "possessive individualism."[1] Large-scale industrial production is carried out by wage workers who sell their labor power to the owners of capital. For such a system to work, laborers must have possession of their own power of work; they must possess themselves and not be the possession of others.

Note, however, that such workers were predominantly male. To work efficiently under these new conditions, the old divisions of labor between men and women needed to be reinforced. Men worked outside the home as productive laborers, women inside as reproductive laborers. Their task was to provide constantly for the male worker the renewal, the re-creation his conditions of work demanded, as well as to rear the next generation of young workers. Only sometimes could women function directly as productive wage laborers in addition to their reproductive work. As the nineteenth century wore on, this division of labor was steadily strengthened. By contrast with feudal society, men were no longer the possession of others; however, if they possessed nothing else, they possessed their women. The social order was not merely capitalist but patriarchal as well.

The third requirement of developing economic relations was presumptive equality for the growing bourgeoisie. Entrepreneurs needed to acquire and dispose of both real and personal property, which required a legal system that would guarantee them redress against nobles and, above all, access to political power. In practice, this was achieved by the supremacy of a parliament of commoners.

The changing mode of production which the emergent capitalist order of the seventeenth century represented demanded solutions to a wholly new range of technical problems. A mercantile and trading society required new and more accurate navigational techniques for merchants' ships, new methods of extraction of raw materials, and new processes of handling these materials when extracted. The techniques for generating solutions to these problems, and the body of knowledge that accumulated as a result of solving them, represented one of the fundamental transitions in the history of humanity, the emergence of modern science, an emergence that can be dated surprisingly precisely to northwestern Europe in the seventeenth century.

The new scientific knowledge, unlike the old precapitalist forms of knowledge, was not passive but active. Whereas in the past philosophers had contemplated the universe, for science in the post-Newtonian world the test of theory was practice, a credo given ideological form by the writings of Francis Bacon. A steady acquisition of facts about the world and its experimental manipulation in the light of those facts were integral to the new theories. No longer was it adequate merely to quote the authority of the ancients; and if the ancient words of wisdom were not in accord with today's observations, they must be discarded. The new science, like the new capitalism, was part of the liberation of humanity from the shackles of feudal serfdom and human ignorance (the links are beautifully displayed in Brecht's *Galileo*). Even the most abstract pronouncements of physics, such as Newton's laws of motion, could be seen as arising out of the social needs of an emergent class.[2] Science was thus an integral part of the new dynamic of capital, even though the fuller articulation of the links between them would take another two centuries to develop.[3]

The Articulation of Bourgeois
Scientific Ideology

It is relatively easy to see the social determinants of science and to show the forces that urge particular problems forward and retard others as thus expressive of social needs as perceived by a dominant class. What is less clear, however, is how the nature of scientific knowledge is itself structured by the social world. And yet some such correspondence must exist. To view the universe and to extract explanatory principles and unifying hypotheses from the rich confusion of phenomena and processes, one must systematize and use tools for systematization that are derived from the experience of the social world and of one's fellow students of the natural world.

It is precisely at this point that the concept of ideology becomes of paramount importance in making transparent the ways in which human understanding becomes refracted by the social order in which that understanding develops. To understand the concerns and modes of explanation of bourgeois science, one must understand the underpinnings of bourgeois ideology.

The radical reorganization of social relations that marked the rise of bourgeois economy had, as a concomitant, the rise of an ideology expressive of these new relations. This ideology, which dominates today, was both a reflection onto the natural world of the social order that was being built and a legitimizing political philosophy by which the new order could be seen as following from eternal principles. Long in advance of the revolutions and regicides of the seventeenth and eighteenth centuries that marked the final triumph of the bourgeois order, intellectuals and political pamphleteers were creating the philosophy to which these revolutions looked for justification and explanation.

It is hardly surprising, then, that the philosophical principles enunciated by the philosophers of the Enlightenment should turn out to be just those that corresponded with the demands of bourgeois social relations. The emphasis of the new bourgeois order on the twin ideas of freedom and equality provided the revolutionary rhetoric of the new class struggling to throw off the grips of church and aristocracy. It was a rhetoric that was to be liberatory, and yet finally, once the victory of the bourgeois class was assured, was to contain within itself the contradictions with which the bourgeois order is faced today.

The eighteenth-century accord between the bourgeois order and its ideology of scientific rationality is typified by the clandestinely published French *Encyclopedia*. Its editor was the physicist and mathematician d'Alembert, and the emphasis throughout was on a secular, rational analysis of both the physical world and human institutions. The motif of scientific rationality, as opposed to the religious themes of faith, the supernatural, and tradition, was obviously a primary requirement for the development of productive forces based on new technological discoveries. Labor, too, had to be reorganized and relocated, in workshops whose productive activities were based on calculations of efficiency and profit, not customary relations. The machine model of the universe gained intellectual hegemony, ceasing to be regarded as merely a metaphor and becoming, instead, the "self-evident" truth about how to look at the world.

The Bourgeois View of Nature

Thus the bourgeois view of nature shaped and was shaped by the science that it developed, organized along certain basic reductionist principles. The rise of modern physics, first with Galileo and then particularly with Newton, ordered and atomized the natural world. Beneath the surface world in all its infinite variety of colors, textures, and varied and transient objects, the new science found another world of absolute masses interacting with one another according to invariant laws that were as regular as clockwork. Causal relationships linked falling bodies, the motion of projectiles, the tides, the moon, and the stars. Gods and spirits were abolished or relegated merely to the "final cause" which set the whole clockwork machinery in motion. (Actually Newton himself remained both religious and mystic throughout his life, but that is one of the minor quirks of personal history: The effect of Newtonian thought was the reverse of Newton's personal philosophy.) The feudal world's universe thus became demystified and, in a manner, disenchanted as well.

This change did not occur without a struggle against those interests that the rising world view opposed. The threat to the Church when astronomers like Copernicus and Galileo sought to replace an earth-centered model of the motion of the heavenly bodies by a sun-centered one was not about cosmology alone, for the Church perceived it as a challenge to a Church-centered world order on earth which mirrored the heavens above. The astronomers, in the spirit of the new capitalism, were challenging heavenly and earthly understanding simultaneously, which was why Bruno, who was most explicit about this, was burned, while Galileo was allowed merely to recant and Copernicus to be published with a little proviso that heliocentrism was merely a theory that made calculations easier but which should not be confused with reality.

In the new world that emerged after Newton, once again heavenly and earthly orders were in seeming harmony. The new physics was dynamic and not static, as were the new processes of trade and exchange. The old world view was replaced with a set of new abstractions in which a series of abstract forces between atomistic and unchanging masses underlay all transactions between bodies. Drop a pound of lead and a pound of feathers from the Leaning Tower of Pisa and the lead will arrive at the ground first because the feathers will be more retarded by air pressure, frictional forces, and so forth. But in Galileo's and Newton's equations the pound of feathers and the pound of lead arrive simultaneously because the *abstract* pound of lead and pound of feathers are equivalent unchanging masses to be inserted into the theoretical equations of the laws of motion.

Sohn-Rethel[4] has pointed out how these abstractions paralleled the world of commodity exchange in which the new capitalism dealt. To each object there are attached properties, mass, or value which are equivalent to or can be exchange for objects of identical mass or value. Commodity exchange is timeless, unmodified by the frictions of the real world; for example, a coin does not change its value by passing from one hand to another, even if it is slightly damaged or worn in the process. Rather, it is an abstract token of a particular exchange value. It was not until the nineteenth century that this view could become fully dominant. The demonstration by Joule that all forms of energy and heat, electromagnetism and chemical reactions were interchangeable and related by a simple constant, the mechanical equivalent of heat (and the later demonstration by Einstein of the equivalence of matter and energy), corresponded to an economic reductionism whereby all human activities could be assessed in terms of their equivalents of pounds, shillings, and pence.*

Humans themselves ceased to be individuals with souls to be saved but became merely hands capable of so many hours of work a day, needing to be stoked with a given quantity of food so that the maximal surplus value could be extracted from their labor. Dickens described that epitome of the nineteenth-century rising capitalist, Thomas Gradgrind of Coketown, as a man

"with a rule and a pair of scales, and the multiplication table always in his pocket, sir, ready to weigh and measure any parcel of human nature, and tell you exactly what it comes to. It is a mere question of figures, a case of simple arithmetic. . . . Time itself for the manufacturer becomes its own machinery: so much material wrought up, so much food consumed, so many powers worn out, so much money made."[5]

*Lest there be any doubt, we should emphasize again that there are two types of criteria for understanding the scientific process. That we can show the social determinants of a particular view of the world, how and why it emerges, says nothing about the truth claims or otherwise of the scientific statements. That Joule's mechanical equivalent of heat or Einstein's matter/energy equivalence were developed in a particular facilitating social framework does not entitle one to conclude that they are thereby by definition either true or false. Criteria for judging the truth of Joule's or Einstein's claims lie between science and the real world, not between science and the social order. We are not committing the "genetic fallacy."

For bourgeois society, nature and humanity itself had become a source of raw materials to be extracted, an alien force to be controlled, tamed, and exploited in the interests of the newly dominant class. The transition from the precapitalist world of nature could not be more complete.[6]

So far we have discussed science in general, or rather physics as though it were all of science. But how did the new mechanical and clockwork vision of the physicists affect the status of living organisms? Just as modern physics starts with Newton, so modern biology must begin with Descartes—philosopher, mathematician, and biological theorist.

In Part V of his *Discourses* of 1637, Descartes analogizes the world, animate and inanimate, to a machine (the *bête machine*). It is this Cartesian machine image that has come to dominate science and to act as the fundamental metaphor legitimating the bourgeois world view, whether of individuals or of the "solid machine" in which they are embedded. That the machine was taken as a model for the living organism and not the reverse is of critical importance. The machine is as much the characteristic symbol of bourgeois productive relations as the "body social" was of feudal society. Bodies are indissoluble wholes that lose their essential characteristics when they are taken into pieces.

Life following life through creatures you dissect,
You lose it in the moment you detect.[7]

Machines, on the contrary, can be disarticulated to be understood and then put back together. Each part serves a separate and analyzable function, and the whole operates in a regular, lawlike manner that can be described by the operation of its separate parts impinging on each other.

Descartes's machine model was soon extended from nonhuman to human organisms. It was clear that many—in fact most—human functions were analogous to those of other animals and therefore were also reducible to mechanics. However, humans had consciousness, self-consciousness, and a mind, which for Descartes, a Catholic, was a soul; and by definition the soul, touched by the breath of God, could not be mere mechanism. So there had to be two sorts of stuff in nature: matter, subject to the mechanical laws of physics; and soul, or mind, a nonmaterial stuff which was the consciousness of the individual, his or her immortal fragment. How did mind and matter interact? By way of a particular region of the brain, Descartes speculated, the pineal gland, in which the mind/soul resided when incorporate, and from which it could turn the knobs, wind the keys, and activate the pumps of the body mechanism.

So developed the inevitable but fatal dysjunction of Western scientific thought, the dogma known in the case of Descartes and his successors as "dualism." As we shall see, some sort of dualism is the inevitable consequence of any sort of reductionist materialism that does not in the end wish to accept that humans are "nothing but" the motion of their molecules. Dualism was a solution to the paradox of mechanism that would enable religion and reductionist science to stave off for another two centuries their inevitable final contention for ideological supremacy. It was a solution compatible with the capitalist order of the day because in weekday affairs it enabled humans to be treated as mere physical mechanisms, objectified and capable of exploitation without contradiction, while on Sundays ideological control could be reinforced by the assertion of the immortality and free will of an unconstrained incorporeal spirit unaffected by the traumas of the workaday world to which its body had been subjected. Today as well, dualism continually reemerges in persistent and various manners from the ashes of the most arid of mechanical materialism.

The Development of a Materialist Biology

For the confident and developing science of the eighteenth and nineteenth centuries, dualism was but a stepping stone toward more thoroughgoing mechanical materialism. Although the analogies changed and became more sophisticated as physical science advanced—from clockwork and hydraulic to electrical and magnetic, and onward to telephone exchanges and computers—the main thrust remained reductionist. For the progressive rationalists of the eighteenth century, science was about cataloguing the states of the world. If a complete specification of all particles at a given time could be achieved, everything would become predictable. The universe was determinate, and the laws of motion applied precisely across a scale ranging from the atoms to the stars. Living organisms were not immune from these laws. The demonstration by Lavoisier that the processes of respiration and the sources of living energy were exactly analogous to those of the burning of a coal fire—the oxidation of foodstuffs in the body tissues —was perhaps the most striking vindication of this approach. It was the first time that a programmatic statement that life must be reducible to molecules could be carried into practice.

But progress in the identification of body chemicals was slow. The demonstration that the substances of which living organisms are composed are only "ordinary," albeit complicated, chemicals came early in the nineteenth century. The intractability of the giant biological molecules—proteins, lipids, nucleic acids—to the available analytical tools remained a stumbling block. The mechanists could make programmatic statements about how life was reducible to chemistry, but these were largely acts of faith. Not until a century after the first nonorganic synthesis of the simple body chemicals did the molecular nature and structures of the giant molecules began to be resolved (and really not until the 1950s did progress became very rapid). The last remaining faith that there would be some special "life force" operating among them which distinguished them absolutely from lesser, nonliving chemicals lingered until the 1920s.[8]

Nonetheless, a radically reductionist program characterized the statements of many of the leading physiologists and biological chemists of the nineteenth century. In 1845 four rising physiologists—Helmholtz, Ludwig, Du Bois Reymond, and Brucke—swore a mutual oath to account for all bodily processes in physiochemical terms.[9] They were followed by others: for instance, Moleschott and Vogt, thoroughgoing mechanical materialists who claimed that humans are what they eat, that genius is a question of phosphorus, and that the brain secretes thought as the kidney secretes urine; and Virchow,* one of the leading figures in the development of cell theory, who was also part of a long

*Virchow's arguments worked both ways: His emphasis on "the body politic" also implied an argument that diseases of individuals were essentially socially caused rather than caused by, for instance, germs. Virchow's emphasis on social medicine, with its progressive and nonreductionist implications, is part of the contradiction between the radical social intent of much of this physiological thought in the nineteenth century and its ultimately repressive ideology.

tradition of social thought which argued that social processes could be described by analogy with the workings of the human body.

It is important to understand the revolutionary intentions of this group. They saw their philosophical commitment to mechanism as a weapon in the struggle against orthodox religion and superstition. Several of them were also militant atheists, social reformers, or even socialists. Science would alleviate the misery of the poor and strengthen the power of the state against the capitalists—and even, to some measure, help democratize society. Their claims were part of the great battle between science and religion in the nineteenth century for supremacy as the dominant ideology of bourgeois society, a fight whose outcome was inevitable but whose final battlefield was not to be physiological reductionism but Darwinian natural selection. The best-known philosopher of the group was Feuerbach, and it was against his version of mechanical materialism that Marx launched his famous theses.[10]

The theses on Feuerbach proved the starting point for Marx's own —and more explicitly Engels's—long-running attempts to transcend mechanical materialism by formulating the principles of a materialist but nonreductionist account of the world and humanity's place within it: dialectical materialism. But within the dominant perspective of biology in the Western tradition, Moleschott's mechanical materialism was to win out, stripped of its millenarian goals and, by the late twentieth century, revealed as an ideology of domination. When today biochemists claim that "a disordered molecule produces a diseased mind,"[11] or psychologists argue that inner city violence can be cured by cutting out sections of the brains of ghetto militants, they are speaking in precisely this Moleschottian tradition.

To complete the mechanical materialist world picture, however, a crucial further step was required: the question of the nature and origin of life itself. The mystery of the relationship of living to nonliving presented a paradox to the early mechanists. If living beings were "merely" chemicals, it should be possible to recreate life from an appropriate physico-chemical mix. Yet one of the biological triumphs of the century was the rigorous demonstration by Pasteur that life only emerged from life; spontaneous generation did not occur. The resolution of this apparent paradox, which had led to many confused polemics between chemical reductionists and the residual school of biological vitalists who continued to oppose them, awaited the Darwinian synthesis, which was able to show that although life came from other living organisms and could not now arise spontaneously, each generation of living things changed, evolved, as a result of the processes of natural selection.

With the theory of evolution came a crucial new element in the understanding of living processes: the dimension of time.[12] Species were not fixed immemorially but were derived in past history from earlier, "simpler" or more "primitive," forms. Trace life back to its evolutionary origins and one could imagine a primordial warm chemical soup in which the crucial chemical reactions could occur. Living forms could coalesce from this prebiotic mix. Darwin speculated about such origins, although the crucial theoretical advances depended on the biochemist Oparin and the biochemical geneticist Haldane in the 1920s (both, incidentally, consciously attempting to work within a dialectical and nonmechanist framework). Experiments only began to catch up with theory from the 1950s onward.

In one sense, evolutionary theory itself represents the apotheosis of a bourgeois world view, just as its subsequent development reflects the contradictions within that world view. The breakdown of the old static feudal order and its replacement with a continually changing and developing capitalism helped introduce the concept of mutability into biology. The age-old daily and seasonal rhythms and the "simple" movement of life from birth through maturity to death had characterized feudalism, but now each generation experienced a world qualitatively different from that of its predecessors. This change was, for the rising eighteenth-century bourgeoisie, progressive. Time's arrow pointed forward irreversibly; it did not loop back upon itself. Understanding of both the earth and life upon it was transformed. Geology slowly came to recognize that the earth had evolved, rivers and seas had moved, layers of rock had been laid down in time sequences atop one another—not in accord with the Biblical myth of creation and the flood but in a steady and uniform sequence over many thousands or millions of years. The principle of uniformitarianism in the hands of such early nineteenth-century geologists as Lyell destroyed the Biblical date for the creation of the earth, 4004 B.C.

And what of life itself? The resemblances and differences of species, their apparent grading virtually one into the other, seemed to imply more than mere coincidence. The discovery of fossils in rock formations whose ages could be estimated implied that some species that had once existed were no longer extant, while new ones had emerged. The doctrine of evolutionism had become an inevitability. At first, in the hands of eighteenth- and early nineteenth-century zoological philosophers like Lamarck and Erasmus Darwin, evolution itself was progressivist, but not in discord with a higher godly design. For Lamarck, species perfected themselves by striving, by modifying their properties to environmental demands and passing these modifications on to their offspring, just as humans were no longer "fixed" in place but could ascend a social hierarchy by virtue—in the liberal myth—of their own efforts. For the elder Darwin evolution was change onward and upward, steadily toward an always more perfect and harmonious future.

It was for Charles Darwin, and the dourer context of the mid-nineteenth century, to frame the mechanisms for evolutionary change in terms of natural selection. Drawing on ideas earlier expressed in the human context by Malthus, he saw that the fact that individuals produced more offspring than survived, and that those better adapted to their environment were more likely to survive long enough to breed in their turn, provides a motor for evolutionary change. Further, Darwinian evolution by natural selection applied not merely to nonhuman species but, it was immediately apparent, to humans as well. It was this observation that set the stage for the final conflict of science with religion, despite the reluctance of many on both sides of the debate to get drawn into it. For, far more than the programmatic statements of the physiological mechanists, Darwinian theory was a direct challenge to the residual hold of Christianity as the dominant ideology of Western society and was seen as such by friend and foe alike.

In retreat since Newton, orthodox Christianity had fallen back into the belief in a God who was first cause of the natural world and still remained the day-to-day controller of life—and especially of human destiny. Darwinism wrested God's final hold on human affairs from his now powerless hands and relegated the deity to, at the best, some dim primordial principle whose will no longer determined human action.

The consequence was to change finally the form of the legitimating ideology of bourgeois society. No longer able to rely upon the myth of a deity who had made all things bright and beautiful and assigned each to his or her estate—the rich ruler in the castle or the poor peasant at the gate—the dominant class dethroned God and replaced him with science. The social order was still to be seen as fixed by forces outside humanity, but now these forces were natural rather than deistic. If anything, this new legitimator of the social order was more formidable than the one it replaced. It has, of course, been with us ever since.

Natural-selection theory and physiological reductionism were explosive and powerful enough statements of a research program to occasion the replacement of one ideology—of God—by another: a mechanical, materialist science. They were, however, at best only programmatic, pointing along a route which they could not yet trace. For example, in the absence of a theory of the gene, Darwinism could not explain the maintenance of inherited variation that was essential for the theory to work. The solution awaited the development of genetic theory at the turn of the twentieth century with the rediscovery of the experiments done by Mendel in the 1860s. This in turn produced the neo-Darwinian synthesis of the 1930s and the recurrent attempts to parcel out biological phenomena into discrete and essentially additive causes, genetic and environmental: the science of biometry.

The Quantification of Behavior

Moleschott's claim that the brain secretes thought as the kidney secretes urine was perhaps the most extreme of the materialist claims of the nineteenth century, but it expresses at the same time the ultimate goal of the philosophy. It was not merely life, but consciousness and human nature itself which must be brought within the reach of rulers, scales, and chemical furnaces. To achieve such a goal it was necessary first to have a theory of behavior, which was no longer seen as a continuous and only partially predictable flow of human action arising from the demands of the soul, of free will, and of the vagaries of human character, stuff for the novelist rather than the scientist; instead, behaviors—now in the plural—had to be seen as a series of discrete and separable units, each of which could be distinguished and analyzed. It was no longer enough to see the body alone as a machine; the role of the brain in organizing and controlling behavior became the center of research attention.

For one school, the brain was an integrative organ whose properties were in some way holistic functions of the entire mass of tissue. For another, these functions were atomized and localized in different regions. This latter was essentially the claim made by the phrenological school of Gall and Spurzheim beginning in Germany and France at the end of the eighteenth century. All human faculties, it claimed, could be broken down into discrete units—abilities such as mathematics or propensities such as love of music or of producing children (philoprogenitiveness).[13] Further, these different abilities and propensities were located in different regions of the brain, and their extent could be assessed from outside by looking at the shape of an individual head or skull. Despite a period of high fashion, phrenology's empirical claims were laughed out of court by the orthodox science of the mid-nineteenth century, but a crucial series of fundamental claims remained intact. These were of the existence of discrete measurable traits that could be localized to specific brain regions. By the end of the nineteenth century the localization school of neuropsychology was clear that different regions of the brain controlled different functions; clear as a result of the postmortem examination of brains of patients whose disabilities had been studied before death; by the somewhat macabre investigations of the behavior of soldiers dying of brain injuries in the battlefields of the Franco-Prussian War; and by experiments with animals. There were brain regions associated with sensory, motor, and association functions; with speech, memory, and affect. It followed that differences in behavior between individuals might be accounted for in terms of differences in structure of different brain regions. There was much dispute as to whether brain size as measured in life by head circumference, or after death by directly weighing, might be associated with intelligence or achievement—an obsession of a number of distinguished nineteenth-century neuroanatomists who anxiously surveyed their colleagues, and left their own brains for analysis by posterity. The systematic distortion of the evidence by nineteenth-century anatomists and anthropologists in attempts to prove that the differences in brain size between male and female brains were biologically meaningful, or that blacks have smaller brains than whites has been devastatingly exposed in a detailed reevaluation by Stephen J. Gould.[14]

The obsession with brain size continued well into the twentieth century. Both Lenin's and Einstein's brains were taken for study after death. Lenin's had an entire institute of brain research founded for its study; years of work could find nothing unusual about the brain, but the institute remains as a major research center. The point is that there are no sensible questions that neuroanatomy can address to the dead brain of however distinguished a scientist or politician.* There is virtually no observable relationship between the size or structure of an individual brain measured after death and any aspect of the intellectual performance of its owner measured during life. There are exceptions: In cases of specific brain damage due to illness, lesions, or tumors, or the brain shrinkage of senile dementia or alcoholism, though even here there are counterexamples.[15] But in general, once the effects of height, age, etc. have been allowed for, brain weight is related to body size. The search for the seat of differences in performance between individuals must move beyond the simple examination of brain structures.

Despite this, there remains a common assumption that there is a relationship between large heads and high brows and intelligence, an assumption that was made the basis of a criminological theory of types by the Italian Cesare Lombroso in the late nineteenth century. According to Lombroso, in an extension of the phrenological theorizing of the early part of the century, criminals could be identified by certain basic physiological features:

The criminal by nature has a feeble cranial capacity, a heavy and developed jaw, projecting [eye] ridges, an abnormal and asymmetrical cranium . . . projecting ears, frequently a crooked or flat nose. Criminals are subject to [color blindness]; left-handedness is common; their muscular force is feeble. . . . Their moral degeneration corresponds with their physical, their criminal tendencies are manifested in infancy by [masturbation], cruelty, inclination to steal, excessive vanity, impulsive character. The criminal by nature is lazy, debauched, cowardly, not susceptible to remorse, without foresight, . . . his handwriting is peculiar . . . his slang is widely diffused. . . . The general . . . persistence of an inferior race type . . .[16]

*Any more than there are useful questions to be asked of the sperm of a septuagenarian Nobel laureate, despite Dr. William Shockley's enthusiasm for donating these fruits of his loins to a "genetic repository" in California where it may be used to inseminate the hopeful bearers of "high IQ" children.

Lombroso and his followers attempted to establish a system whereby a predisposition to engage in antisocial behavior could be predicted on the basis of physical characteristics; from surveys conducted in prisons he concluded among other things that murderers have "cold, glassy, blood-shot eyes, curly abundant hair, strong jaws, long ears and thin lips"; forgers are "pale and amiable, with small eyes and large noses; they become bald and grey-haired early"; and sex criminals have "glinting eyes, strong jaws, thick lips, lots of hair and projecting ears."[17]

A rational criminology thus became possible, a theory of criminal faces that was the obvious forerunner to today's belief in criminal chromosomes. The strength of Lombroso's typology is that it drew from current myths about the criminal and gave them apparent scientific support. The myths found their way routinely into mass culture, as in Agatha Christie, for instance. In an early book we find her clean-cut young upper-class English hero secretly observing the arrival of a Communist trade unionist at a rendezvous: "The man who came up the staircase with a soft-footed tread was quite unknown to Tommy. He was obviously of the very dregs of society. The low beetling brows, and the criminal jaw, the bestiality of the whole countenance were new to the young man, though he was a type that Scotland Yard would have recognised at a glance."[18] Lombroso would have recognized him too.

Implicit in such criminology is the belief that individual behaviors can be located as the fixed properties of individuals, as characteristic as their height or hair color. Also implicit within the research program that such a reductionist biological determinism maintains is the claim that it is possible to compare the behaviors of different individuals across some appropriate scale. Behaviors are not all-or-none. Like height, they are continuously distributed variables; individual A is more aggressive than individual B, or less so than C. If one could devise appropriate scales, like rulers for height, one should be able to plot the distribution of the entire population on a scale for aggression, criminality, or whatever. It is the belief in such a distribution that provides the rationale for thinking about IQ tests as measures of intelligence, which is discussed in Chapter 5. If all individuals within a population can be placed, for any particular trait, along a linear distribution, the famous bell-shaped "normal" curve is produced. Individuals who fall outside the majority portion of this distribution are abnormal, or deviant.

Because we take the concept of deviance so easily for granted, because it seems so "natural," it is important to remember how recently it has appeared in the history of bourgeois society. The concepts of criminality, madness, and indeed illness itself—their treatment by seclusion, in prisons, asylums, and hospitals—only developed slowly from the seventeenth century and with accelerating pace through the nineteenth century.[19] It is not that prior to the bourgeois revolution there was no theory of human nature. Typological theory argued that human temperament was fixed as a sort of titration of the four basic types—phlegmatic, bilious, choleric, and sanguine. Concepts of the fixity of human evil and of original sin clashed with the possibility of redemption through faith or good works. Certainly criminal codes existed, as did madness and disease. But medieval and early capitalist society tolerated a far greater range of human variation than was to be acceptable later. Peddlers and vagabonds, rogues and eccentrics were part of life's stage: consider the characters in a Breughel or Hogarth painting, or an eighteenth-century picaresque novel. The reductionist materialism of the nineteenth century sought to control, regularize, and limit this variation. Or think of the transition between the multitudinous richness of characters of an early Dickens novel like *Pickwick Papers* and the later accounts of the conformity of the new bourgeoisie portrayed in *Dombey and Son* or *Hard Times*. The social institutions of an industrial society could decreasingly tolerate deviance, which became a meaningful concept only when there was a norm, a concept of the average, from which people could be argued to deviate.*

The Origin of Behavior

Behaviors, then, in the reductionist view, may be quantified, distributed in relationship to a norm, or located in some way "in the

---

*Indeed, in writing this book we have become aware of the extent to which there are still large cross-cultural differences in how norms are viewed. The U.S. educational system, it appears to us, is far more interested in categorizing the children that pass through it as "within the normal range" or alternatively as deviant from it; American parents are more likely to be told that their child falls "outside the norm" than in England, where perhaps a greater range of behavior in children is taken for granted —or less is expected of them.

brain." But how do they arise? This too was a major concern of nineteenth-century theorizing. We have shown how the inheritance of behavior, of human nature, forms a major theme of the Victorian novelists from Disraeli to Dickens and Zola. The theory that behaviors, even trivial ones, are inherited rather than acquired was clearly articulated by Charles Darwin in his book *The Expression of Emotion in Man and Animals*. In it, for instance he notes:

A gentleman of considerable position was found by his wife to have the peculiar trick, when he lay fast asleep on his back in bed, of raising his right arm slowly in front of his face up to his forehead, then dropping it with a jerk so that the wrist fell heavily on the bridge of the nose. . . . Many years after his death his son married a lady who had never heard of the family incident. She however, observed precisely the same peculiarity of her husband, but his nose not being particularly prominent has never as yet suffered from a blow. . . . One of his children, a girl, has inherited the same trick.[20]*

While Darwin was collecting anecdotes, Galton was measuring, quantifying, and attempting to define the laws of ancestral inheritance of such behaviors. The inheritance or otherwise of such foibles as Darwin records was not, of course, the central question. In genetic studies from Darwin's day to the present, most of the attention directed to human behavior has been concerned with two major themes: the inheritance of intelligence and the inheritance of mental illness or criminality. One of the major purposes of the collection of psychometric evidence (to be discussed in relation to IQ in Chapter 5) was to measure the degree to which any given behavior was inherited rather than environmentally shaped. The spurious dichotomization of nature and nurture begins here.

While the techniques used in *Hereditary Genius*[21] were crude, the questions asked and the methodology developed soon after were to

_____

*How would *we* explain such an anecdote? For us, it is analogous to some of the stories of amazing coincidences among separated identical twins popular today—or the search for explanations for ESP, UFOs, and bent spoons. We begin by being skeptical of the phenomena. And we point out that scientific research and explanation are concerned above all with the understanding of regularities and repeatable phenomena, not exceptions and flukes, many of which, like the apparent coincidence of behavior of long-separated identical twins, simply disappear on closer analysis.

remain practically unchanged for the century dividing Darwin and Galton from the modern generation of biological determinists. The sorry history of this century of insistence on the iron nature of biological determination of criminality and degeneracy, leading to the growth of the eugenics movement, sterilization laws, and the race science of Nazi Germany has frequently been told.[22] It is not our purpose here to retrace that history. Rather, we are concerned with the way in which the philosophy of reductionism, and its intimate intertwining with biological determinism, developed into the modern synthesis of sociobiology and molecular biology.

The Central Dogma:
The Core of the Mechanistic Program

The nineteenth-century themes of the chemicalization of physiology, the quantification of behavior, and the genetic theory of evolution would have remained only programmatic insights without the explosive growth of biological theory and method of the last thirty years. To substantiate them turned out to require more than slogans and mathematics. What were needed were the powerful new machines and techniques for the determination of the structure of the giant molecules, for observing the microscopic internal structure of the cells, and, above all, for studying the dynamic interplay of individual molecules within the cell. By the 1950s it had begun to be possible to describe and account for, in the mechanistic sense, the behavior of individual body organs—muscles, liver, kidneys, etc.—in terms of the properties and interchange of individual molecules: the mechanist's dream.

The grand unification between the concerns of the geneticists and those of the mechanistic physiologists came in the 1950s—the "crowning triumph" of twentieth-century biology, the elucidation of the genetic code. This required a theoretical addition to the mechanistic program, to be sure. Hitherto it had been sufficient to claim that a full accounting for the biological universe and the human condition was possible by an understanding of the trio of *composition*—the molecules the organism contains; *structure*—the ways these molecules are arranged in space; and *dynamics*—the chemical interchanges among the molecules. To this now needed to be added a fourth concept, that of *information*.

The concept of information itself had an interesting history, arising as it did from attempts during the Second World War to devise guided missile systems, and, through the 1950s and 1960s, in laying the theoretical infrastructure for the computer and electronics industries. The understanding that one could view systems and their actions in terms not merely of matter and the energy flow through it but in terms of information exchanges—that molecular structures could convey instructions or information one to the other—shook up a theoretical kaleidoscope, and in one sense made possible Crick, Watson, and Wilkins's recognition that the double helical structure of the DNA molecule could also carry genetic instructions across the generations. Molecules, the energetic interchanges between them, and the information they carried provided the mechanists' ultimate triumph, expressed in Crick's deliberate formulation of what he called the "central dogma" of the new molecular biology: "DNA → RNA → protein."[23]* In other words, there is a one-way flow of information between these molecules, a flow that gives historical and ontological primacy to the hereditary molecule. It is this that underlies the sociobiologists' "selfish gene" arguments that, after all, the organism is merely DNA's way of making another DNA molecule; that everything, in a preformationist sense that runs like a chain through several centuries of reductionism, is in the gene.

It is hard to overemphasize the ideological organizing function that this type of formulation of the mechanics of the transcription of DNA into protein fulfills. Long before Crick, the imagery of the biochemistry of the cell had been that of the factory, where functions were specialized for the conversion of energy into particular products and which had its own part to play in the economy of the organism as a whole. Some ten years before Crick's formulation, Fritz Lipmann, discoverer of one of the key molecules engaged in energy exchange within the body, ATP, formulated his central metaphor in almost pre-Keynesian economic terms: ATP was the body's energy currency.

*For Crick, "Once information has passed into protein it cannot get out again." For Monod, "One must regard the total organism as the ultimate epigenetic expression of the genetic message itself."[24]

Produced in particular cellular regions, it was placed in an "energy bank" in which it was maintained in two forms, those of "current account" and "deposit account." Ultimately, the cell's and the body's energy books must balance by an appropriate mix of monetary and fiscal policies.[25]

Crick's metaphor was more appropriate to the sophisticated economies of the 1960s in which considerations of production were diminishing relative to those of its control and management. It was to this new world that information theory, with its control cycles, feedback and feed-forward loops, and regulatory mechanisms, was so appropriate; and it is in this new way that the molecular biologists conceive of the cell—an assembly-line factory in which the DNA blueprints are interpreted and raw materials fabricated to produce the protein end products in response to a series of regulated requirements. Read any introductory textbook to the new molecular biology and you will find these metaphors as a central part of the cellular description. Even the drawings of the protein synthesis sequence itself are often deliberately laid out in "assembly line" style. And the metaphor does not merely dominate the teaching of the new biology: It and language derived from it are key features of the way molecular biologists themselves conceive of and describe their own experimental programs.

And not merely molecular biologists. The synthesis of physiology and genetics that an information theory containing a double helix provided was steadily extended upward from individuals to populations and their origins. The integrated reductionist world views presented by biologically determinist writings like those of E. O. Wilson (*Sociobiology: The New Synthesis*) or Richard Dawkins (*The Selfish Gene*) draw explicitly on molecular biology's central dogma to define their commitment to the claim that the gene is ontologically prior to the individual and the individual to society,* and equally explicitly on a set of transferred economic concepts developed in the management of the increasingly complex capitalist societies of the sixties and seventies: Cost benefit analysis, investment opportunity costs, game theory, system engineering and communica-

*For Jacques Monod, "You have an exact logical equivalence between these two—the family and the cells. This effect is entirely written in the structure of the protein which is itself written in the DNA."[27]

tion, and the like are all unabashedly transferred into the natural domain.

Drawn from inspection of the human social order, they define sociobiology's world view, and, as we should expect and as happened with Darwinism earlier, they are then reflected back as a justifier for that social order—as, for instance, when economists describe monetarist theories as in accord with the biological condition of humanity.[26] We shall see this process amply exemplified in the chapters that follow. For now, we want only to emphasize how the very transparency and explicitness of Crick's formulation of the "central dogma," and his quasi-religious choice of language in which to cast it, seize and restate the essential ideological concern of this mechanist tradition.

For the mechanical materialists the grand program begun by Descartes has now in its broad outline been completed. All that remains is the filling in of details. Even for the workings of so complex a system as the human brain and consciousness, the end is in sight. Immense amounts are known about the chemical composition and cellular structures of the brain, about the electrical properties of its individual units, and indeed of great masses of brain tissue functioning in harmony. We know how the analyzer cells of the visual system or the withdrawal reflex of a slug given an electric shock may be wired up, and about regions of the brain whose function is concerned with anger, fear, hunger, sexual appetite, or sleep. The mechanists' claims here are clear. In the nineteenth century, Darwin's supporter T. H. Huxley dismissed the mind as no more than the whistle of the steam train, an irrelevant spin-off of physiological function. Pavlov, in discovering the conditioned reflex, believed he had the key to the reduction of psychology to physiology, and one strand of reductionism has followed his lead. In this tradition, molecules and cellular activity cause behavior, and, as genes cause molecules, the chain that runs from particular unusual genes, say, to criminal violence and schizophrenia is unbroken.

Much of what follows in this book will be an explanation of the inadequacy of the claims for these causal chains, both on theoretical and on empirical grounds, as well as an analysis of their ideological roles in the defense of biological determinist views of the human condition. Only then can we move on to show how these reductionist models may be transcended by a biology more fully in accord with the reality and complexity of the material world. Before that, however, we must examine the contradictions of the other twin planks of bourgeois ideology: the necessity for freedom and equality in the social domain. To do this, we must retrace our steps to the emergence of bourgeois society from feudalism.

CHAPTER FOUR

THE LEGITIMATION OF INEQUALITY

The process of change from feudal to bourgeois society was marked, from its inception in the fourteenth century and with increasing intensity after the seventeenth century, by constant conflict and struggle. Just as Roman and feudal societies were repeatedly upset by servile revolts like the slave uprisings of Spartacus and Nat Turner, or the peasant revolts in Germany and Russia, so bourgeois society has been marked by incidents such as the rick burning and machine breaking of Captain Swing in nineteenth-century Britain, and the patriarchy reinforced by periodic episodes of witch-hunting. So too the last decades have been marked by uprisings: of blacks in America, of workers in Poland, of unemployed youth in Britain. The pattern is similar in each case: at all times the violence of those who do not have against those who do is close to realization, and when it erupts it is met by the organized police power of the state. Yet it is an obvious disadvantage to those with power to have to meet violence with violence. The outcomes of violent confrontations are not always sure. They may spread; property and wealth are destroyed; production is disrupted; and the tranquility of the possessors to enjoy the fruits of their possessions is disturbed. It is clearly better if the struggle can be moved to an institutional level—the courts, the parliamentary process, the negotiating table. Since these institutions are themselves in the hands of the possessors of social power, the outcome is better assured, and, if concessions must be made for fear of successful disruption, those concessions can be small, slow, and even illusory. Those who have power must, if possible, avoid the struggle entirely, or at least keep it in bounds that can be accommodated within the institutions that they control. To do either requires the weapon of ideology. Those who possess power and their representatives can most effectively disarm those who would struggle against them by convincing them of the legitimacy and inevitability of the reigning social organization. If what exists is right, then one ought not oppose it; if it exists inevitably, one can never oppose it successfully.

Up to the seventeenth century, the main propagator of legitimacy and inevitability was the Church, through the doctrine of grace and divine right. Even Luther, the religious rebel, commanded that the peasants obey their lord. Moreover, he stood clearly for order: "Peace is more important than justice; and peace was not made for the sake of justice, but justice for the sake of peace."[1] To the degree that ideological weapons have been successful in convincing people of the justice and inevitability of present social arrangements, any attempt to revolutionize society must use ideological counterweapons that deprive the old order of its legitimacy and at the same time build a case for the new.

The Contradictions

The change in social relations brought about by the bourgeois revolution required more than simply a commitment to rationality and science. The need for freedom and equality of individuals—to move geographically, to own their own labor power, and to enter into a variety of economic relations—was supported by a commitment to individual freedom and equality as absolute, God-given rights—at least to males. The French *Encyclopedia* was not merely a rationalist technical work. Diderot, Voltaire, Montesquieu, Rousseau, and its other contributors made of the *Encyclopedia* a manifesto of political liberalism that matched its scientific rationalism. The hundred years from Locke's *Two Treatises on Civil Government*, which justified the English Revolution, to Paine's *Rights of Man*, which justified the French, was the period of invention and elaboration of an ideology of freedom and equality that was claimed to be unchallengeable. "We hold these truths to be self-evident," the composers of the American Declaration of Independence wrote, "that all men are created equal, that they are endowed by their Creator with certain inalienable rights, that among these rights are life, liberty and the pursuit of happiness" (i.e. wealth).

Yet, when the framers of the Declaration of Independence wrote that "all men are created equal," they meant quite literally "men," since women certainly did not enjoy these rights in the new republic. However, they did not mean literally "all men" since black slavery continued after both the American and the French revolutions. Despite the universal and transcendental terms in which the manifestos of the revolutionary bourgeoisie were framed, the societies that were being built were much more restricted. What was required was equality of merchants, manufacturers, lawyers, and tax farmers with the formerly privileged nobility, not the equality of all persons. The freedom needed was the freedom to invest, to buy and sell both goods and labor, to set up shop in any place and at any time without the hindrance of feudal restrictions on commerce and labor, and to possess women as reproductive labor. What was not needed was the freedom of all human beings to pursue happiness. As in Orwell's *Animal Farm*, all were equal, but some more equal than others.

The problem in creating an ideological justification is that the principle may prove rather more sweeping than the practice demands. The founders of liberal democracy needed an ideology to justify and legitimate the victory of the bourgeoisie over the entrenched aristocracy, of one class over another, rather than an ideology that would eliminate classes and patriarchy. Yet they also needed the support of the *menu peuple*, the yeoman farmers, and the peasants, in their struggle. One can hardly imagine making a revolution with the battle cry "Liberty and justice for some!" So the ideology outstrips the reality. The pamphleteers of the bourgeois revolution created, by necessity and no doubt in part by conviction, a set of philosophical principles in contradiction with the social reality they intended to build.

The final victory of the bourgeoisie over the old order meant that the ideas of freedom and equality that had been the subversive weapons of a revolutionary class now became the legitimating ideology of the class in power. The problem was and still is that the society created by the Revolution was in obvious contrast with the ideology from which it drew its claims of right. Slavery continued in French St. Dominique until the successful slave revolt of 1801 and in Martinique for a further fifty years. It was abolished only in 1833 in British dominions and not until 1863 in the United States. Suffrage, even among the free, was greatly restricted. After the Reform Bill of 1832 in Britain still only about 10 percent of the adult population was enfranchised, and not until 1918 was universal manhood suffrage established. Woman suffrage waited until 1920 in the United States, 1928 in Britain, 1946 in Belgium, and 1981 in Switzerland. The rights of women to own property and to enter any job they choose on a par with men was and still remains a battleground.

More fundamentally, economic and social power remain extremely unequally distributed and show no sign of being effectively redistributed. Despite the idea of equality, some people have power over their own lives and the lives of others, while most do not. There remain rich people and poor people, employers who own and control the means of production and employees who do not even control the conditions of their own labor. By and large, men are more powerful than women and whites more powerful than blacks. The income distribution in the United States and Britain is clearly unequal, with about 20 percent of the income accruing to the highest 5 percent of the families and only 5 percent accruing to the lowest-paid 20 percent. The distribution of wealth is much more skewed. The richest 5 percent own 50 percent of all the wealth in the United States, and if one discounts the houses people live in, the cars they drive, and the clothes they wear, then nearly all wealth belongs to the richest 5 percent.[2]*

*For example, 1 percent own 60 percent of all corporate stock, and the wealthiest 5 percent own 83 percent of stock.

Nor can a case be made that economic equality has dramatically increased over the last three hundred years. Using the admittedly rough figures gathered by Gregory King in 1688 from hearth taxes,[3] one can estimate that at the time of the Glorious Revolution the poorest 20 percent of families had 4 percent of the income and the richest 5 percent received 32 percent of the income. The income distribution has become somewhat more equal during the last hundred years, but the figures are based on money income. In the United States, for example, the proportion of the work force in agriculture has dropped from 40 percent to 4 percent, so no account is taken of the loss of real income as the poorest groups have moved out of subsistence agriculture. On the other hand, there have been periodic expansions of poor law and welfare payments which have had the effect of redistributing income, but these have fluctuated considerably. It would be extremely difficult to show that the industrial working poor at the height of the Chartist movement in the 1840s were better off than their rural ancestors of Tudor times, and there is considerable evidence that the early part of the nineteenth century saw greater misery for the poor.[4] Even the redistribution of income that has occurred in the last hundred years has hardly had the effect of creating a society of equality. In the United States the infant mortality rate among blacks is 1.8 times that for whites, and the average expectation of life is 10 percent lower.[5] In Britain, perinatal mortality is more than twice as high for infants born to working-class families as it is to families of professionals.[6]

Political ideology may separate people on the question of the origins, morality, and future of economic and social inequality, but no one can question its existence. Bourgeois society, like the aristocratic feudal society it replaced, is characterized by immense differences in status, wealth, and power. The fact that there has been growth in the economy over time, so that in every generation—at least until the present —children are better off than their parents, and that there have been great shifts in the labor force—from a production to a service economy, for instance—serves merely to mask these differences.

The perpetual struggle between those who possess power and those over whom they exercise that power is exacerbated in bourgeois society by a contradiction between ideology and reality that did not apply in feudal times. The political ideology of freedom and, especially, equality that legitimated the overthrow of the aristocracy helped to produce a society in which the idea of equality is still as subversive as ever, if taken seriously. It is in the name of equality and the end of injustice that the Paris Commune of 1871, the student/worker uprisings of 1968, and the uprisings of blacks in the inner cities of Britain and America have taken place. Clearly, if the society in which we live is to seem just, both to the possessors and to the dispossessed, some different understanding of freedom and equality is needed, one that brings the reality of social life into congruence with the moral imperatives. It is precisely to meet the need for self-justification and to prevent social disorder that the ideology of biological determinism has been developed.

Dealing with the Contradictions:
The Three Claims of Biological Determinism

The ideology of equality has become transformed into a weapon in support of, rather than against, a society of inequality by relocating the cause of inequality from the structure of society to the nature of individuals. First, it is asserted that the inequalities in society are a direct and ineluctable consequence of the differences in intrinsic merit and ability among individuals. Anyone may succeed, get to the top; but whether one does so or not is a consequence of inherent strength or weakness of will or character. Second, while liberal ideology has followed a cultural determinism, emphasizing circumstance and education, biological determinism locates such successes and failures of the will and character as coded, in large part, in an individual's genes; merit and ability will be passed from generation to generation within families. Finally, it is claimed that the presence of such biological differences between individuals of necessity leads to the creation of hierarchical societies because it is part of biologically determined human nature to form hierarchies of status, wealth, and power. All three elements are necessary for a complete justification of present social arrangements.

The determinative role of individual difference in molding the structure of modern bourgeois society has been made quite explicit. Lester

Frank Ward, a major figure in nineteenth-century American sociology, wrote that education is

the power that is destined to overthrow every species of hierarchy. It is destined to remove all artificial inequality and leave the natural inequalities to find their true level. The true value of a newborn infant lies . . . in its naked capacity for acquiring the ability to do.[7]

The concept was given up-to-date form by the English sociologist Michael Young in the 1960s in his satire *The Rise of the Meritocracy.*[8] This meritocracy was soon to be given biological underpinnings. By 1969 Arthur Jensen of the University of California could claim in his article on IQ and achievement:

We have to face it, the assortment of persons into occupational roles simply is not "fair" in any absolute sense. The best we can hope for is that true merit, given equality of opportunity, act as a basis for the natural assorting power.[9]

Lest the political consequences of this natural inequality escape us, some determinists draw them out quite explicitly. Richard Herrnstein of Harvard, one of the most active ideologues of meritocracy, explains that:

The privileged classes of the past were probably not much superior biologically to the downtrodden, which is why revolution had a fair chance of success. By removing artificial barriers between classes, society has encouraged the creation of biological barriers. When people can take their natural level in society, the upper classes will, by definition, have greater capacity than the lower.[10]

Here the scheme of explanation is laid out in its most explicit form. The *ancien régime* was characterized by artifical barriers to social movement. What the bourgeois revolutions did was to destroy those arbitrary distinctions and allow natural differences to assert themselves. Equality, then, is equality of opportunity, not equality of ability or result. Life is like a foot-race. In the bad old days the aristocrats got a head start (or were declared the winners by fiat), but now everyone starts together so that the best win—best being determined biologically. In this scheme, society is seen as composed of freely moving individuals, social atoms, who, unimpeded by artifical social conventions, rise or fall in the social hierarchy in accordance with their desires and innate abilities. Social mobility is completely open and fair, or may require at most a minor adjustment, an occasional regulatory act of legislation, to make it so. Such a society has naturally produced about as much equality as is possible. Any remaining differences constitute the irreducible minimum of inequality, engendered by natural differences in true merit. The bourgeois revolutions succeeded because they were only breaking down artificial barriers, but new revolutions are futile because we cannot eliminate natural barriers. It is not quite clear what principle of biology guarantees that biologically "inferior" groups cannot seize power from biologically "superior" ones, but it is clearly implied that some general property of stability accompanies "natural" hierarchies.

By putting this gloss on the idea of equality, biological determinism converts it from a subversive to a legitimizing ideal and a means of social control. The differences in society are both fair and inevitable because they are natural. Thus, it is both physically impossible to change the status quo in any thoroughgoing way and morally wrong to try.

A political corollary of this view of society is a prescription for the activity of the state. The social program of the state should not be directed toward an "unnatural" equalization of social condition, which in any case would be impossible because of its "artificiality," but rather the state should provide the lubricant to ease and promote the movement of individuals into the positions to which their intrinsic natures have predisposed them. Laws promoting equal opportunity are to be encouraged, but artificial quotas that guarantee, say, 10 percent of all jobs in some industry to blacks are wrong because they attempt to reduce inequality below its "natural" level. In like manner, rather than give the same education to blacks and whites or to working-class and upper-middle-class children, schools should sort them by IQ tests or "eleven-plus" examinations into their appropriate "natural" educational environments. Education in fact becomes the chief institution for promoting social assortment according to innate ability. "The power that is destined to overthrow every species of hierarchy" is "universal education."[11]

The second, crucial, step in the building of the ideology of biological determinism, following the claim that social inequality is based on intrinsic individual differences, is the equation of *intrinsic* with *genetic*. It is possible, in principle, for differences in individuals to be inborn without being biologically heritable. Indeed, explanations of inequality based on individual successes or failures of will or character often seek to go no further. In fact from the biological perspective a considerable proportion of the subtle physiological and morphological variation between individuals in strains of experimental animals can be shown to be the result of accidents of development that are not heritable. Nor does the everyday understanding of inborn differences necessarily equate them with what is inherited. The conflation of intrinsic and inherited qualities is a distinct step in building the structure of biological determinism.

The theory that we live in a society that rewards intrinsic merit is at variance with common observation in one important respect. It is evident that parents in some way pass on their social power to their children. The sons of oil magnates tend to become bankers, while the children of oil workers tend to be in debt to banks.* The probability that any of the Rockefeller brothers would have spent their lives working in a Standard Oil garage is fairly small. While there is certainly considerable social mobility, the correlation between social status of parents and children is high. The often quoted study of the American occupational structure by Blau and Duncan, for example, showed that 71 percent of the sons of white-collar workers were themselves white-collar workers, while 62 percent of the sons of blue-collar workers remained in the blue-collar category.[12] The British figures are not dissimilar. Such figures, however, vastly underestimate the degree of fixity of social class, since most movement between white- and blue-

*This correlation was first pointed out in the nineteenth century by Francis Galton, the inventor of a host of anthropometric techniques for quantifying aspects of human performance. Galton is the progenitor of intelligence measuring techniques and of theories of its hereditary nature. In 1869, in his book *Hereditary Genius*, he traced the family trees of a large number of eminent Victorian bishops, judges, scientists, etc., and by showing that their fathers and grandfathers had also tended to be bishops, judges, scientists, and so forth, he concluded comfortingly that genius was inherited and that it was concentrated disproportionately among Victorian upper-class males. Other classes in Britain and other nationalities in Europe possessed a lesser quantity of genius, and the nonwhite "races" least of all.

collar categories is horizontal with respect to income, status, control of working conditions, and security. The nature of particular jobs changes between generations. There are fewer workers in primary production and more in service industries today. Yet clerks are no less proletarian because they sit at desks rather than standing at benches; and salespersons, one of the largest groups of "white-collar workers," are among the lowest paid and least secure of all occupational groups. Can it be that parents pass on their social status to their children in defiance of the meritocratic process? Unless bourgeois society has, like its aristocratic predecessor, artificial inherited privilege, the passage of social power from parents to children must be natural. Differences in merit are not only intrinsic but biologically inherited: They are in the genes.

The convergence of the two meanings of inheritance—the social and the biological—legitimizes the passage of social power from generation to generation. It can still be asserted that we have an equal opportunity society with each individual rising or falling in the social scale according to merit, provided we understand that merit is carried in the genes. The notion of the inheritance of human behavior and therefore of social position which so permeated the literature of the nineteenth century can thus be understood not as an intellectual atavism, a throwback to aristocratic ideas in a bourgeois world, but, on the contrary, as a consistently worked out position to explain the facts of bourgeois society.

The claim of inherited differences in merit and ability between individuals does not complete the argument for the justice and inevitability of bourgeois social arrangements. There remain logical difficulties that must be coped with by the determinist. First, there is the naturalistic fallacy that draws "ought" from "is." Whether or not there are biological differences between individuals does not, in itself, provide a basis for what is "fair." Ideas of justice cannot be derived from the facts of nature, although, of course, one may begin with the a priori that what is natural is good—provided one is willing to accept, for example, that the blinding of infants by trachoma is "just." Second, there is the equation of "innate" with "unchangeable," which seems to imply some dominance of the natural over the artificial. Yet the history of the human species is precisely the history of social victories over nature, of mountains moved, seas joined, diseases eradicated, and even species made over for human purposes. To say that all these have been done "in accordance with the laws of nature" is to say nothing more than that we live in a material world with certain constraints. But what those constraints are must be determined in each case. "Natural" is not "fixed." Nature can be changed according to nature.

These are not simply formal objections to determinism. They have political force. Intrinsic differences between individuals in ability to perform social functions have not always been regarded as leading necessarily to a hierarchical society. Marx summed up his vision of Communist society in the "Critique of the Gotha Programme" as "From each according to his abilities, to each according to his need." In the 1930s geneticists like J. B. S. Haldane, who was a member of the British Communist Party and a columnist for the *Daily Worker*, and H. J. Muller, who worked in the Soviet Union after the Bolshevik Revolution and who, at the time, identified himself as a Marxist, argued (along lines that we would not) that important aspects of human behavior were influenced by genes.[13] Yet both believed that social relations could be revolutionized and classes abolished despite individual intrinsic differences. Social democrats and liberals have expressed the same idea. One of the leading evolutionists of the twentieth century, Theodosius Dobzhansky, argued in *Genetic Diversity and Human Equality*[14] that we may build a society in which picture painters and house painters, barbers and surgeons, can receive equal psychic and material rewards, although he believed them to differ genetically from each other.

It seems that the simple assertion that there are inherited differences in ability between individuals has been insufficient to justify the continuation of a hierarchical society. It must be further claimed that those heritable differences necessarily and justly lead to a society of differential power and reward. This is the role played by human nature theories, the third of the components to the biological determinists' claims. In addition to the biological differences said to exist between individuals or groups, it is supposed that there are biological "tendencies" shared by all human beings and their societies, and that these tendencies result in hierarchically organized societies in which individuals

compete for the limited resources allocated to their role sector. The best and most entrepreneurial of the role-actors usually gain a disproportionate share of the rewards, while the least successful are displaced to other, less desirable positions.[15]

The claim that "human nature" guarantees that inherited differences between individuals and groups will be translated into a hierarchy of status, wealth, and power completes the total ideology of biological determinism. To justify their original ascent to power, the new middle class had to demand a society in which "intrinsic merit" could be rewarded. To maintain their position they now claim that intrinsic merit, once free to assert itself, *will* be rewarded, for it is "human nature" to form hierarchies of power and reward.

On Human Nature

The appeal to "human nature" has been characteristic of all political philosophies. Hobbes claimed that the state of nature was "the war of all against all," but Locke, to the contrary, saw tolerance and reason as the human natural state. Social Darwinism took "nature red in tooth and claw" as the primitive state for humans, while Kropotkin claimed cooperation and mutual aid as basic to human nature. Even Marx, whose historical and dialectical materialism are hostile to a fixity of human nature, took the basic nature of the human species to be the transforming of the world to satisfy its own needs. For Marx, one realized one's humanity in labor.

Biological determinism, as we have been describing it, draws its human nature ideology largely from Hobbes and the social Darwinists, since these are the principles on which bourgeois political economy are founded. In its most modern avatar, sociobiology, the Hobbesian ideology even derives cooperation and altruism, which it recognizes as overt characteristics of human social organization, from an underlying competitive mechanism. Sociobiology, drawing its principles directly from Darwinian natural selection, claims that tribalism, entrepreneurial activity, xenophobia, male domination, and social stratification are dictated by the human genotype as molded during the course of evolution. It makes the two assertions, inevitability and justice, that are required if it is to serve as a legitimization and perpetuation of the social order. Thus E. O. Wilson writes in *Sociobiology*:

If the planned society—the creation of which seems inevitable in the coming century—were to deliberately steer its members past those stresses and conflicts that once gave the destructive phenotypes their Darwinian edge, the other phenotypes might dwindle with them. In this, the ultimate genetic sense, social control would rob man of his humanity.[16]

Before trying to plan society, then, we must await the most definite knowledge about the human genotype. Moreover, "A genetically accurate and hence [sic] completely fair code of ethics must also wait."[17]

Cultural Reductionism?

Critics of the biological determinist position are frequently challenged as to the alternatives that they espouse. While we must emphasize that posing such alternatives is not required to expose the fallacies in an argument, we would nevertheless like to accept that challenge here. But we should make clear the framework within which we accept it. When biological determinists discuss their critics, they tend to label them as "radical environmentalists," that is, they oppose biological determinism by arguing that it is possible to divorce an understanding of the human condition and of human differences entirely from biology. There are indeed schools of thought that have argued in this way. We are not among them. We must insist that a full understanding of the human condition demands an integration of the biological and the social in which neither is given primacy or ontological priority over the other but in which they are seen as being related in a dialectical manner, a manner that distinguishes epistemologically between levels of explanation relating to the individual and levels relating to the social without collapsing one into the other or denying the existence of either. Nonetheless, we must look briefly at some of the major modes of culturally reductionist thinking and the fallacies that underlie them. They may be grouped into two types. The first gives ontological primacy to the social over the individual, and is hence the complete antithesis of biological determinism. The second, while reinstating the individual against the social, does so as if the individual had no biology at all.

The first type of cultural reductionism is exemplified by certain tendencies in "vulgar" Marxism, in sociological relativism, and in antipsychiatry and deviancy theory. Vulgar Marxism is a form of economic reductionism that locates all forms of human consciousness, knowledge, and cultural expression as determined by the mode of economic production and the social relations that this engenders. Knowledge of the natural world, then, is no more than an ideology that expresses an individual's class position relative to the means of production, and it changes as the economic order changes. Individuals are ultimately shaped in all but the most trivial ways by their social circumstances: the iron laws of economic history determine a historically infinitely plastic "human nature" and mechanically cause human actions. Disease, illness, depression, and the pain of day-to-day living are no more than the inevitable consequence of a capitalist and patriarchal social order. The only "science" is economics. This type of reductionism, which discounts human consciousness as a mere epiphenomenon of the economy, is of course in a strange way a close relative of social Darwinism: one finds its expression in the line of social and political writing that runs from Kautsky through to some contemporary Trotskyist theorists (for instance, Ernest Mandel[18]) on the left.

Against this economic reduction as the explanatory principle underlying all human behavior, we would counterpose the understanding of Marxist philosophers like Georg Lukacs[19] and Agnes Heller,[20] and of revolutionary practitioners and theorists like Mao Tse-tung[21] on the power of human consciousness in both interpreting and changing the world, a power based on an understanding of the essential dialectical unity of the biological and the social, not as two distinct spheres, or separable components of action, but as ontologically coterminous.

The bourgeois manifestation of economic reductionism takes the form of a cultural pluralism which maintains that all forms of human actions or belief are determined by "interest." The "reality" of the natural world is subordinate to beliefs about it, and there is no way of adjudicating between the claims to truth made by one group of scientists compared with those made by another. What Wilson, Dawkins, or Trivers write about sociobiology reflects their interests in advancing their own social position. What we write reflects ours. We and they may be the objects of anthropological inquiry by sociologists of knowledge whose own position relative to "truth" seems strangely unaffected, though where they find the rock on which to stand among these quicksands of "interest" seems unclear. The most explicit formulation of this "science as social relations" argument may be found in, for instance, the writings of the Edinburgh historians, sociologists, and philosophers of science—Barnes, Bloor, and Shapin.[22]

How this kind of theoretical position works out in practice may be seen in the strong development of a sociological theory of deviancy and of antipsychiatry over the last two decades. For these cultural reductionists, individual behavior does not exist except as a consequence of social labeling. While the biological determinist sees a child's unruly behavior in school as commanded by his or her genes, ghetto violence as caused by abnormal molecules in the brains of "ringleaders," or male dominance in society as part of evolutionary survival mechanisms, deviancy theory dissolves away all such phenomena as mere labels. A child is labeled as "stupid," a schizophrenic is labeled as "mad" because society needs to create scapegoats.[23] The cure is then merely to relabel the child, or the schizophrenic, and sweetness and light will flow. The famous account of child relabeling, "Pygmalion in the Classroom"[24] in which children's IQ scores were improved by telling teachers that they are "late developers" and the Laingian approach to the interpretation of schizophrenia both flow from such a viewpoint. Individuals are again infinitely malleable, defined merely as the products of the expectations of their society, and have no separate existence. Their own ontological status and their own biological nature have been dissolved away. Without in any way wishing to deny the importance of labeling as helping shape social interactions and individual's definitions of themselves, again we would insist that a child's classroom performance is not merely the result of what his or her teachers think; a schizophrenic person's existential despair and irrational behavior are not merely the result of being labeled as mad by his or her family or doctors.

The second kind of cultural reductionism we wish to refer to is one in which explanations of behavior are still sought at the level of the individual, but an individual who is nonetheless regarded as biologically empty, a sort of cultural *tabula rasa* on which early experience may mark what it pleases and on which biology has no influence. The later developments of such an individual are then seen as largely determined by such early experiences. Like biological determinism, this sort of reductionism ends up by blaming the victim, but victims are now made by culture rather than biology.

Part of this approach is centered upon individual psychology, part on cultural anthropology and sociology. In psychology, the approach is through psychometry, a procedure that relies heavily on measurement of responses of people to questionnaires and performance of simple tasks and an elaborate array of statistical procedures. Human action itself is reduced to individual reified lumps objectified in the black box of the head. With Spearman, Burt, and Eysenck, the argument runs that intelligence, for example, is a unitary lump; with Guilford, it can be broken down into 120 different factors. The procedures in both are analogous. The elusive dynamic of human action, purposes, intentions, and interrelations is nailed down in multiple correlations of mathematical elegance and biological vacuity. The measurement of this black box is theorized by behaviorism, a school that dominated American psychology from the 1930s to the 1960s, into a system in which specified inputs are connected to specified outputs, and which can change its behavior adaptively, that is, learn in response to contingencies of reinforcement, of reward and punishment. The apparent extreme environmentalism of this school, which developed around Watson and later B. F. Skinner, serves merely to hide its impoverished concept of humanity and its manipulative approach to the control of individual humans, evidenced by Skinner's concern with the control and manipulation of behavior, in children or prisoners, by a superior cadre of value-free demigods in white coats, who are to decide on the correct behavior into which they will coerce their victims.[25] The novel and film *A Clockwork Orange* portrayed one possible consequence of this mode of thinking about and treating human beings. The reality, witnessed in numerous correctional institutions throughout the United States, the notorious Behavior Control units in British prisons, in institutions for the "educationally subnormal," and in the thinking of many schoolteachers trained on a version of the theory, may yet approach such fiction.

In cultural sociology and anthropology, cultural reductionism is embedded in theories that postulate ethnic and class subcultures propagated across generations by purely cultural connections, and which provide different patterns of success and failure for their members. The "culture of poverty" is an example. The poor are characterized by the demand for immediate gratification, by short-term planning, by violence, and by unstable family structure. These characteristics, because they are maladaptive in bourgeois society, doom the poor to continued poverty; and the children of the poor, being so acculturated, cannot escape from the cycle. This theory of the cycle of deprivation has been explicitly espoused by Sir Keith Joseph, one of the key ideologues of the Thatcher government in Britain.* His eugenicist concerns have led him to use the cultural argument rather than a genetic one to support a policy recommendation of easing contraceptive availability for the poor. (A similar conclusion was arrived at from a more explicitly·genetic point of view in the 1930s by the architect of the British welfare state, Lord Beveridge, who argued that if poverty ran in the genes, sterilization for workers on the dole would help eliminate it.)

Extending their scope from the "culturally deprived" to the successfully upward-mobile, determinists explain the disproportionate representation in the United States of Jews among professionals, and particularly among academics, by pointing to a cultural tradition that places emphasis on scholarship, as well as the need for a base of occupational expertise as a hedge against the economic consequences of anti-Semitism. The recent appearance of large numbers of people of Japanese and Chinese ancestry among professionals is given a similar explanation.

Because they are unable to appeal to physical principles as a mechanical basis for cultural inheritance, cultural reductionists are thought of as representing a "soft" science, or even humanistic speculation, and their legitimacy is under attack from the "hard" biological determinists (who themselves, of course, are at the "soft" end of the scale of natural scientific texture). But this kind of cultural reductionism suffers from another, more damaging, softness as an underpinning for political action. If inherited social inequalities are the result of ineluctable biological differences, then the elimination of inequality requires that we change peoples' genes. On the other hand, such individually based, liberal cultural reductionism only requires that we change their heads, or the way others think about them. Hence where others would seek a change in political structure,

*The failure of Britain's Social Science Research Council to commission research that could "prove" Sir Keith's theory correct is widely regarded as one reason for his attempts to abolish the council during his term as Thatcher's minister of education.

such individually based liberal cultural reductionism often places its faith in general and uniform education.

Unfortunately for this belief, however, the immense equalization of education that has occurred in the last eighty years has not been matched by a great equalization of society. In 1900 only 6.3 percent of the 17-year-old population of the United States were high school graduates, while at present it is about 75 percent, yet the unequal distribution of wealth and social power remains.* Indeed, cultural reductionism is directly under attack for the apparent wholesale failure of public education to destroy the class structure. The motivation for Arthur Jensen's 1969 article on IQ in the *Harvard Educational Review*, which signaled the renewed surge of biological determinism, was provided in its opening sentence: "Compensatory education has been tried and has failed." Whether or not compensatory education has really been tried, and whether or not it has failed, it seems likely that if every person in the Western world could read and understand Kant's *Critique of Pure Reason* the ranks of the unemployed would not ipso facto be decreased—though they would be more literate.

Cultural reductionism of this individual kind shares with biological determinism an assumption that the proportion of persons in given roles and with given status in society is determined by the availability of talents and abilities. That is, the demand for, say, doctors, is infinite, and only the paucity of talent available to fill this role controls the number of physicians. In fact, the reverse seems to be true: the number of persons filling particular jobs is determined by structural relations that are almost independent of the potential "supply." If only bankers had children, there would be no change in the number of bankers, although both biological determinism and cultural reductionism predict the contrary.

We have argued that the development of bourgeois society has generated both a serious contradiction and a mode for coping with it. The contradiction is between the ideology of freedom and equality and the actual social dynamic that generates powerlessness and inequality. The mode of coping with that contradiction is a reductionist natural

*A seminar by a well-known French sociologist was once held with the remarkable title, "Why is a better-educated France as unequal as ever?" That is indeed a problem for cultural, not for biological, determinists, who would claim it as evidence for their view.

science that develops simple models of social or biological causation, providing fundamentally flawed explanations of social reality.

The contradiction appears in varied contexts: inequalities between social classes, races, the sexes, the appearance of social deviance. In each case a variant of reductionist, biological determinist theory has been constructed to deal in detail with the specific issue. Once the mode of explanation is set—"there's a gene for it"—the program of research and theory follows for the entire range of individual and social phenomena from autism to the "zero-sum society." In what follows, we examine in detail those forms of the contradiction and the attempts to resolve it that are current and politically vital. That examination is meant not only to reveal the specific errors of the cases in point, but to provide a model for demystifying the inevitable future uses to which biological determinist arguments will be put.

CHAPTER FIVE

IQ: THE RANK ORDERING OF THE WORLD

The Roots of IQ Testing

Social power runs in families. The probability that a child will grow into an adult in the highest 10 percent of income earners is ten times greater for children whose parents were in the top 10 percent than for children of the lowest 10 percent.[1] In France, the school failure rate of working-class children is four times that for children of the professional class.[2] How are we to explain hereditary differences in social power in a society that claims to have abolished hereditary privilege in the eighteenth century? One explanation—that hereditary privilege is integral to bourgeois society, which is not structurally conducive to real equality—is too disquieting and threatening; it breeds disorder and discontent; it leads to urban riots like those in Watts and Brixton. The alternative is to suppose that the successful possess an intrinsic merit, a merit that runs in the blood: Hereditary privilege becomes simply the ineluctable consequence of inherited ability. This is the explanation offered by the mental testing movement, whose basic argument can be summarized in a set of six propositions that, taken as a whole, form a seemingly logical explanation of social inequality. These are:

1. There are differences in status, wealth, and power.
2. These differences are consequences of different intrinsic ability, especially different "intelligence."
3. IQ tests are instruments that measure this intrinsic ability.
4. Differences in intelligence are largely the result of genetic differences between individuals.
5. Because they are the result of genetic differences, differences in ability are fixed and unchangeable.
6. Because most of the differences between individuals in ability are genetic, the differences between races and classes are also genetic and unchangeable.

While the argument begins with an undoubted truth that demands explanation, the rest is a mixture of factual errors and conceptual misunderstandings of elementary biology.

The purposes of Alfred Binet, who in 1905 published the first intelligence test, seem to have been entirely benign. The practical problem to which Binet addressed himself was to devise a brief testing procedure that could be used to help identify children who, as matters then stood, could not profit from instruction in the regular public schools of Paris. The problem with such children, Binet reasoned, was that their "intelligence" had failed to develop properly. The intelligence test was to be used as a diagnostic instrument. When the test had located a child with deficient intelligence, the next step was to increase the intelligence of such a child. That could be done, in Binet's view, with appropriate courses in "mental orthopedics." The important point is that Binet did not for a moment suggest that his test was a measure of some "fixed" or "innate" characteristic of the child. To those who asserted that the intelligence of an individual is a fixed quantity that one cannot augment, Binet's response was clear: "We must protest and react against this brutal pessimism."[3]

The basic principle of Binet's test was extraordinarily simple. With the assumption that the children to be tested had all shared a similar cultural background, Binet argued that older children should be able to perform mental tasks that younger children could not. To put matters very simply, we do not expect the average 3-year-old to be able to recite the names of the months, but we do expect a normal 10-year-old to be able to do so. Thus, a 10-year-old who cannot recite the months is probably not very intelligent, while a 3-year-old who can do so is probably highly intelligent. What Binet did, quite simply, was to put together sets of "intellectual" tasks appropriate for each age of childhood. There were, for example, some tasks that the average 8-year-old could pass, but which were too difficult for the average 7-year-old and very easy for the average 9-year-old. Those tasks defined the "mental age" of eight years. The intelligence of a child depended upon the relation his or her mental and chronological ages bore to each other. The child whose mental age was higher than his or her chronological age was "bright" or accelerated, and the child whose mental age was lower than his or her chronological age was "dull" or retarded. For most children, of course, the mental and chronological ages were the same. To Binet's satisfaction, the mental ages of children in a school class, as measured by his test, tended to correspond with teachers' judgments about which children were more or less "intelligent." That is scarcely surprising, since for the most part Binet's test involved materials and methods of approach similar to those emphasized in the school system. When a child lagged behind its age-mates by as much as two years of mental age, it seemed obvious to Binet that remedial intervention was called for. When two Belgian investigators reported that the children whom they had studied had much higher mental ages than the Paris children studied by Binet, Binet noted that the Belgian children attended a private school and came from the upper social classes. The small class sizes in the private school, plus the kind of training given in a "cultured" home, could explain, in Binet's view, the higher intelligence of the Belgian children.

The translators and importers of Binet's test, both in the United States and in England, tended to share a common ideology, one dramatically at variance with Binet's. They asserted that the intelligence test measured an innate and unchangeable quantity, fixed by genetic inheritance. When Binet died prematurely in 1911, the Galtonian eugenicists took clear control of the mental testing movement in the English-speaking countries and carried their determinist principles even further. The differences in measured intelligence not just between individuals but between social classes and races were now asserted to be of genetic origin. The test was no longer regarded as a diagnostic instrument, helpful to educators, but could identify the genetically (and incurably) defective, those whose uncontrolled breeding posed a "menace . . . to the social, economic and moral welfare of the state."[4] When Lewis Terman introduced the Stanford-Binet test to the United States in 1916 he wrote that a low level of intelligence

is very common among Spanish-Indian and Mexican families of the Southwest and also among negroes. Their dullness seems to be racial, or at least inherent in the family stocks from which they come. . . . The writer predicts that . . . there will be discovered enormously significant racial differences in general intelligence, differences which cannot be wiped out by any scheme of mental culture.

Children of this group should be segregated in special classes. . . . They cannot master abstractions, but they can often be made efficient workers. . . . There is no possibility at present of convincing society that they should not be allowed to reproduce, although from a eugenic point of view they constitute a grave problem because of their unusually prolific breeding.[5]

Though Terman's Stanford-Binet test was basically a translation of Binet's French items, it contained two significant modifications. First, a set of items said to measure the intelligence of adults was included, as well as items for children of different ages. Second, the ratio between mental and chronological age, the "intelligence quotient," or IQ, was now calculated to replace the simple statement of mental and chronological ages. The clear implication was that the IQ, fixed by the genes, remained constant throughout the individual's life. "The fixed character of mental levels" was cited by another translator of Binet's test, Henry Goddard, in a 1919 lecture at Princeton University, as the reason why some were rich and others poor, some employed and other unemployed. "How can there be such a thing as social equality with this wide range of mental capacity? . . . As for an equal distribution of the wealth of the world, that is equally absurd."[6]

The major translator of Binet's test in England was Cyril Burt, whose links to Galtonian eugenics were even more pronounced than those of his American contemporaries. Burt's father was a physician who treated Galton, and Galton's strong recommendations hastened Burt's appointment as the first school psychologist in the English-speaking world. As early as 1909 Burt had administered some crude tests to two very small groups of schoolchildren in the town of Oxford. The children at one school were the sons of Oxford dons, fellows of the Royal Society, etc., while children at the other school were the sons of ordinary townspeople. Burt claimed that the children from the higher-class school did better on his tests and that this demonstrated that intelligence was inherited. This scientifically stated conclusion, published in the 1909 *British Journal of Psychology*,[7] might have been predicted from Burt's handwritten entry, six years earlier, in his Oxford undergraduate notebook: "The problem of the very poor—chronic poverty: Little prospect of the solution of the problem without the forcible detention of the wreckage of society or other preventing them from propagating their species."

Burt continued his eugenic researches into the inheritance of IQ until he died in 1971, knighted by his monarch and bemedaled by the American Psychological Association. The masses of data that he published helped to establish the "eleven-plus" examination in England, linked to the postwar system of selective education. "Intelligence," Burt wrote in 1947, "will enter into everything the child says, thinks, does or attempts, both while he is at school and later on. . . . If intelligence is innate, the child's degree of intelligence is permanently limited." Further, "Capacity must obviously limit content. It is impossible for a pint jug to hold more than a pint of milk; and it is equally impossible for a child's educational attainments to rise higher than his educable capacity permits."[8] There could be no clearer statement of what had happened to Binet's test in the hands of the Galtonians. The test designed to alert educators that they must intervene with special educational treatment was now said to measure "educable capacity." When a child did poorly in school, or when an adult was unemployed, it was because he or she was genetically inferior and must always remain so. The fault was not in the school or in the society, but in the inferior person.

The IQ test, in practice, has been used both in the United States and England to shunt vast numbers of working-class and minority children onto inferior and dead-end educational tracks.* The reactionary impact of the test, however, has extended far beyond the classroom. The testing

---

*"Tracking" in the U.S. educational system is more or less synonymous with "streaming" in Britain.

movement was clearly linked, in the United States, to the passage, beginning in 1907, of compulsory sterilization laws aimed at genetically inferior "degenerates." The categories detailed included, in different states, criminals, idiots, imbeciles, epileptics, rapists, lunatics, drunkards, drug fiends, syphilitics, moral and sexual perverts, and "diseased and degenerate persons." The sterilization laws, explicitly declared constitutional by the U.S. Supreme Court in 1927, established as a matter of legal fact the core assertion of biological determinism: that all these degenerate characteristics were transmitted through the genes. When the IQ testing program of the United States Army in World War I indicated that immigrants from Southern and Eastern Europe had low test scores, this was said to demonstrate that "Alpines" and "Mediterraneans" were genetically inferior to "Nordics." The army IQ data figured prominently in the public and congressional debates over the Immigration Act of 1924. That overtly racist act established as a feature of American immigration policy a system of "national origin quotas." The purpose of the quotas was explicitly to debar, as much as possible, the genetically inferior peoples of Southern and Eastern Europe, while encouraging "Nordic" immigration from northern and western Europe. This tale has been told in full elsewhere.[9]

Today many (if not most) psychologists recognize that differences in IQ between various races and/or ethnic groups cannot be interpreted as having a genetic basis. The obvious fact is that human races and populations differ in their cultural environments and experiences, no less than in their gene pools. There is thus no reason to attribute average score differences between groups to genetic factors, particularly since it is so obviously the case that the ability to answer the kinds of questions asked by IQ testers depends heavily on one's past experience. Thus, during World War I, the Army Alpha test asked Polish, Italian, and Jewish immigrants to identify the product manufactured by Smith & Wesson and to give the nicknames of professional baseball teams. For immigrants who could not speak English, the Army Beta test was designed as a "nonverbal" measure of "innate intelligence." That test asked the immigrants to point out what was missing from each of a set of drawings. The set included a drawing of a tennis court, with the net missing. The immigrant who could not answer such a question was thereby shown to be genetically inferior to the tennis-playing psychologists who devised such tests for adults.

What IQ Tests Measure

How do we know that IQ tests measure "intelligence"? Somehow, when the tests are created, there must exist a prior criterion of intelligence against which the results of the tests can be compared. People who are generally considered "intelligent" must rate high and those who are obviously "stupid" must do badly or the test will be rejected. Binet's original test, and its adaptations into English, were constructed to correspond to teachers' and psychologists' a priori notions of intelligence. Especially in the hands of Terman and Burt, they were tinkered with and standardized so that they became consistent predictors of school performance. Test items that differentiated boys from girls, for example, were removed, since the tests were not meant to make that distinction; differences between social classes, or between ethnic groups or races, however, have not been massaged away, precisely because it is these differences that the tests are *meant* to measure.

IQ tests at present vary considerably in their form and content, but all of them are validated by how well they agree with older standards. It must be remembered that an IQ test is published and distributed by a publishing company as a commercial item, selling hundreds of thousands of copies. The chief selling point of such tests, as announced in their advertising, is their excellent agreement with the results of the Stanford-Binet test. Most combine tests of vocabulary, numerical reasoning, analogical reasoning, and pattern recognition. Some are filled with specific and overt cultural referents: Children are asked to identify characters from English literature ("Who was Wilkins Micawber?"); they are asked to make class judgments ("Which of the five persons below is most like a carpenter, plumber, and a bricklayer? 1) postman, 2) lawyer, 3) truck driver, 4) doctor, 5) painter"); they are asked to judge socially acceptable behavior ("What should you do when you notice you will be late to school?"); they are asked to judge social stereotypes ("Which is prettier?" when given the choice between a girl with some Negroid features and a doll-like European face); they are asked to define obscure words (sudorific, homunculus, parterre). Of course, the "right" answers to such questions are good predictors of school performance.

Other tests are "nonverbal" and consist of picture explanations or geometric pattern recognition. All—and most especially the nonverbal tests—depend upon the tested person having learned the ability to spend long periods participating in a contentless, contextless mental exercise under the supervision of authority and under the implied threat of reward or punishment that accompanies all tests of any nature. Again, they necessarily predict school performance, since they mimic the content and circumstances of schoolwork.

IQ tests, then, have not been designed from the principles of some general theory of intelligence and subsequently shown to be independently a predictor of social success. On the contrary, they have been empirically adjusted and standardized to correlate well with school performance, while the notion that they measure "intelligence" is added on with no independent justification to validate them. Indeed, we do not know what that mysterious quality "intelligence" is. At least one psychologist, E. G. Boring, has defined it as "what intelligence tests measure."[10] The empirical fact is that there exist tests that predict reasonably well how children will perform in school. That these tests advertise themselves as "intelligence" measures should not delude us into investing them with more meaning than they have.

Reifying Behavior

The possibility of behavioral measurements rests upon certain basic underlying assumptions, which should now be clarified. First, it is assumed that it is possible to define, absolutely or operationally, a particular "quality" to be measured. Some such qualities, like height, are relatively unproblematic. To the question "How tall are you?" the answer in centimeters, feet, or inches is easy to give. To the question "How angry are you?" no such easy answer can be given. Anger has to be defined operationally, as, for instance, how often an individual placed in a given test situation and asked the question by the experimenter responds by hitting him on the nose. This is not a flippant example. "Aggression" in a rat is measured by putting a mouse in a cage with it and observing the behavior and time taken for the rat to kill the mouse. Sometimes this is described under the name "muricidal" behavior in the literature, which presumably makes the experimenters happier that they are measuring something really scientific.

Research in this area thus becomes forced into Boring's circularity: Intelligence "is" what intelligence tests measure.

The "quality" is then taken to be an underlying object that is merely reflected in varying aspects of an individual's behavior under widely different circumstances. Thus "aggression" is what individuals express when a man beats his wife, pickets boycott scabs during a strike, teenagers fight after a football game, black Africans struggle for independence from their colonial masters, generals press buttons unleashing thermonuclear war, or America and the Soviet Union compete in the Olympic Games or the space race. The underlying quality is identical with that which underlies muricide in rats.

Second, it is assumed that the quality is a fixed property of an individual. Aggression and intelligence are seen not as processes that emerge from a situation and are part of the relationships of that situation, but rather exist like reservoirs each of defined amount, inside each of us, to be turned on or off. Instead of seeing the anger or aggression expressed in inner city riots as emerging from the interaction between individuals and their social and economic circumstances and as expressive of collective action—therefore a social phenomenon—the biological determinist argument defines inner city violence as merely the sum of individual units of aggressiveness. So psychosurgeons like Mark and Ervin call for a program of research to find and cure the physical lesions that cause urban ghetto riots (see Chapter 7).

Thus verbs are redefined as nouns; processes of interaction are reified and located inside the individual. Further, reified verbs, like aggression, are assumed to be rigid, fixed things that can be reproducibly measured. Like height, they will not vary much from day to day; indeed, if the tests designed to measure them show such variations they are regarded as poor tests. It is assumed not that the "quality" being measured is labile, but that our instruments need greater precision.

Psychometry and the
Obsession with the Norm

Implicit in reification is the third and crucial premise of the mental testing movement. If processes are really things that are the properties of individuals and that can be measured by invariant objective rules, then there must be scales on which they can be located. The scale must be metric in some manner, and it must be possible to compare individuals across the scale. If one person has an aggression score of 100 and the next of 120, the second is therefore 20 percent more aggressive than the first. The fault in the logic should be clear: The fact that it is possible to devise tests on which individuals score arbitrary points does not mean that the quality being measured by the test is really metric. The illusion is provided by the scale. Height is metric, but consider, for instance, color. We could present individuals with a set of colors ranging from red to blue and ask them to rank them as 1 (reddest) to 10 (bluest). But this would not mean that the color rated 2 was actually twice as blue as the color rated 1. The ordinal scale is an arbitrary one, and most psychometric tests are actually ordinals of this sort. If one rat kills ten mice in five minutes, and a second rat kills twelve in the same time, this does not automatically mean that the second is 20 percent more aggressive than the first. If one student scores 80 in an exam and a second 40, this does not mean the first is twice as intelligent as the second.

Surmounting or disguising the scaling problem is integral to the grand illusion of psychometry. Individuals vary in height, and if heights for a hundred or so individuals drawn at random from a population are plotted, they will likely fall into the normal distribution, or bell-shaped curve. If the divisions in one's scale are very fine—say, inches—the bell curve is quite wide. If we had no measures less than feet, and we measured each individual to the nearest foot, the curve would be much narrower at the bottom. The vast majority of individuals in Western society would lie between the five- and six-foot measure. While we know the relationship of inches to feet and could under appropriate circumstances convert from one scale to another, and we know when to use each, as when we are finding a pair of shoes that fit or deciding the best size to make a door opening, we do not know the comparable relationships between different ways of measuring aggression or intelligence. Which scale is chosen depends on whether one wants to make differences of scale appear large or small, and these decisions are those that psychometry arbitrarily makes. The decision that a "good" scale is one in which two-thirds of the population should lie within 15 percent of the mean score of the entire population—the famous normal distribution—is arbitrary, but its power is such that psychometrists chop and change their scales till they meet this criterion.

Yet the power of the "norm," once established, is that it is used to judge individuals who have been located along its linear scale. Deviations from the norm are regarded with alarm. Parents who are told that their child is two standard deviations from the norm on some behavioral score are led to believe that he or she is "abnormal" and should be adjusted in some way to psychometry's Procrustean bed. Psychometry, above all, is a tool of a conformist society that, for all its professed concern with individuals, is in reality mainly concerned to match them against others and to attempt to adjust them to conformity.

Pressure to conform to social norms, and institutions that propagate and reinforce these norms, are, of course, characteristic of all human societies. In advanced capitalist societies and today's state capitalist societies like the Soviet Union or those in Eastern Europe, the norm becomes an ideological weapon in its own right, foreshadowed by Huxley's *Brave New World* and Orwell's *1984* but cloaked in the benign language of those who only wish to help, to advise, but not to control and manipulate. Let us be clear: norms are statistical artifacts; they are not biological realities. Biology is not committed to bell-shaped curves.

IQ Tests as Predictors of Social Success

The claim that IQ tests are good predictors of eventual social success is, except in a trivial and misleading sense, simply incorrect. It is true that if one measures social success by income or by what sociologists call socioeconomic status (SES)—a combination of income, years of schooling, and occupation—then people with higher incomes or higher SES did better on IQ tests when they were children than did people with low incomes or low SES. For example, a person whose childhood IQ was in the top 10 percent of all children is fifty times more likely to wind up in the top 10 percent of income than a child whose IQ was in the lowest 10 percent. But that is not really quite the question of interest. What we really should ask is: How much more likely is a high-IQ child to wind up in the top 10 percent of income, *all other things being equal?* In other words, there are multiple and complex causes of events which do not act or exist independently of each other. Even where *A* looks at first sight as if it is the cause of *B*, it sometimes really turns out on deeper examination that *A* and *B* are both effects of some prior cause, *C.* For example, on a worldwide basis, there is a strong positive relationship between how much fat and how much protein the population of a particular country consumes. Rich countries consume a lot of each, poor countries little. But fat consumption is neither the cause nor the result of eating protein. Both are the consequence of how much money people have to spend on food. Thus, although fat consumption per capita is statistically a predictor of protein consumption per capita, it is not a predictor when all other things are equal. Countries that have the same per capita income show no particular relation between average fat and average protein consumption, since the real causal variable, income, is not varying between countries.

This is precisely the situation for IQ performance and eventual social success. They go together because both are the consequences of other causes. To see this, we can ask how good a predictor IQ is of eventual economic success when we hold constant the person's family background and the number of years of schooling. With these constant, a child in the top 10 percent of IQ has only twice, not fifty times, the chance of winding up in the top 10 percent of income as a child of the lowest IQ group. Conversely, and more important, a child whose family is in the top 10 percent of economic success has a 25 times greater chance of also being at the top than the child of the poorest 10 percent of families, even when both children have average IQ.[11] Family background, rather than IQ, is the overwhelming reason why an individual ends up with a higher than average income. Strong performance on IQ tests is simply a reflection of a certain kind of family environment, and once that latter variable is held constant, IQ becomes only a weak predictor of economic success. If there is indeed an intrinsic ability that leads to success, IQ tests do not measure it. If IQ tests do measure intrinsic intelligence as is claimed, then clearly it is better to be born rich than smart.

The Heritability of IQ

The next step in the determinist argument is to claim that differences between individuals in their IQ arise from differences in their genes. The notion that intelligence is hereditary is, of course, deeply built into the theory of IQ testing itself because of its commitment to the measurement of something that is intrinsic and unchangeable. From the very beginning of the American and British mental testing movement, it was assumed that IQ was biologically heritable.

There are certain erroneous senses of "heritable" that appear in the psychometricans' writings on IQ, mixed up with the geneticists' technical meaning of heritability, and which contribute to false conclusions about the consequences of heritability. The first error is that genes themselves determine intelligence. Neither for IQ nor for any other trait can genes be said to determine the organism. There is no one-to-one correspondence between the genes inherited from one's parents and one's height, weight, metabolic rate, sickness, health, or any other nontrivial organic characteristic. The critical distinction in biology is between the *phenotype* of an organism, which may be taken to mean the total of its morphological, physiological, and behavioral properties, and its *genotype*, the state of its genes. It is the genotype, not the phenotype, that is inherited. The genotype is fixed; the phenotype develops and changes constantly. The organism itself is at every stage the consequence of a developmental process that occurs in some historical sequence of environments. At every instant in development (and development goes on until death) the next step is a consequence of the organism's present biological state, which includes both its genes and the physical and social environment in which it finds itself. This comprises the first principle of developmental genetics: that every organism is the unique product of the interaction between genes and environment at every stage of life. While this is a textbook principle of biology, it has been widely ignored in determinist writings. "In the actual race of life, which is not to get ahead, but to get ahead of somebody," wrote E. L. Thorndike, the leading psychologist of the first half of the century, "the chief determining factor is heredity."[12]

The second error—even if admitting that genes do not determine the actual developmental outcome—is to claim that they determine the effective limit to which it can go. Burt's metaphor of the pint jug that can hold no more than a pint of milk is a precise image of this view of genes as the determinants of capacity. If the genetic capacity is large, the argument runs, then an enriched environment will result in a superior organism, although in a poor environment the same individual will not show much ability. If the genetic capacity is poor, however, then an enriched environment will be wasted. Like the notion of the absolute determination of organisms by genes, this view of genetic "capacity" is simply false. There is nothing in our knowledge of the action of genes that suggests differential total capacity. In theory, of course, there must be *some* maximum height, say, to which an individual could grow; but in fact there is no relationship between that purely theoretical maximum, which is never reached in practice, and the actual variations among individuals. The lack of relationship between actual state and theoretical maximum is a consequence of the fact that growth rates and growth maxima are not related. Sometimes it is the slowest growers that reach the greatest size. The proper description of the difference between genetic types is not in some hypothetical "capacity" but in the specific phenotype that will develop for that genotype as a consequence of some specific chain of environmental circumstances.

Nor, of course does the phenotype develop linearly from the genotype from birth to adulthood. The "intelligence" of an infant is not merely a certain small percentage of that of the adult it will become, as if the "pint jug" were being steadily filled. The process of growing up is not a linear progression from incompetence to competence: To survive, a newborn baby must be competent at being a newborn baby, not at being a tiny version of the adult it will later become. Development is not just a quantitative process but one in which there are transformations in quality—between suckling and chewing solid food, for instance, or between sensorimotor and cognitive behavior. But such transitions are not permitted in the rank-ordered view of the universe that determinism offers.

The total variation in phenotype in a population of individuals arises from two interacting sources. First, individuals with the same genes still differ from each other in phenotype because they have experienced different developmental environments. Second, there are different genotypes in the population which differ from each other on the average even in the same array of environments. The phenotype of an individual cannot be broken down into the separate contributions of genotype and of environment, because the two interact to produce the organism; but the total variation of any phenotype in the population can be broken down into the variation between the average of the different genotypes and the variation among individuals with the same genotype. The variation between the average performance of different genotypes is called the *genetic variance* of the trait (that is, the aspect of the phenotype under study—eye color, height, or whatever) in the population, while the variation among individuals of the same genotype is called the *environmental variance* of the trait in the population. It is important to notice that the genetic and environmental variances are not universal properties of a trait but depend upon which population of individuals is being characterized and under which set of environments. Some populations may have a lot of genetic variance for a character, some only a little. Some environments are more variable than others.

The *heritability* of a trait, in the technical sense in which geneticists understand it, is the proportion of all the variation of a trait in a population that is accounted for by the genetic variance. Symbolically,

$$\text{Heritability} = H = \frac{\text{genetic variance}}{\text{genetic variance} + \text{environmental variance}}$$

If the heritability is 100 percent, then all of the variance in the population is genetic. Each genotype would be phenotypically different, but there would be no developmental variation among individuals of the same genotype. If the heritability is zero, all of the variation is among individuals within a genotype, and there is no average variation from genotype to genotype. Characters like height, weight, shape, metabolic activity, and behavioral traits all have heritabilities below 100 percent. Some, like specific language spoken or religious and political affiliation, have heritabilities of zero. The claim of biological determinists has been that the heritability of IQ is about 80 percent. How do they arrive at this figure?

Estimating the Heritability of IQ

All genetic studies are studies of the resemblances of relatives. If a trait is heritable, that is, if different genotypes have different average performances, then relatives ought to resemble each other more closely than unrelated persons do, since relatives share genes from common ancestors. Brothers and sisters ought to be more like each other than aunts and nephews, who ought to be more similar than totally unrelated people. The standard measure of similarity between things that vary quantitatively is their *correlation*, which measures the degree to which larger values for one variable go together with larger values of a second variable, and smaller values with smaller values. The correlation coefficient, r, ranges from $+1.0$ for perfect positive correlation, through zero for no relationship, to -1.0 for perfect negative correlation. So, for example, there is a positive correlation between father's income and child's years of schooling. Richer fathers have better-educated children while poorer fathers have less-educated children, on the average. The correlation is not perfect, since some poor families produce children who go to graduate school, but it is positive. In contrast, in the United States there is a negative correlation between family income and the number of visits per year to hospital emergency rooms. The lower your income, the more likely you are to use the emergency room as a medical service instead of a private doctor.

One important point about correlation is that it measures how two things vary together but does not measure how similar their average levels are. So the correlation between the heights of mothers and their sons could be perfect in that taller mothers had the taller sons and shorter mothers had the shorter sons, yet all the sons could be taller than all the mothers. Covariation is not the same as identity. The significance of this fact for the heritability of IQ and its meaning is considerable. Suppose a group of fathers had IQs of 96, 97, 98, 99, 100, 101, 102, and 103, while their daughters, separated from their fathers at birth and raised by foster parents, had IQs respectively of 106, 107, 108, 109, 110, 111, 112, and 113. There is a perfect correspondence between the IQs of fathers and daughters, and we might judge the character to be perfectly heritable because, knowing a father's IQ, we could tell without error which of the daughters was his. The correlation is, in fact,

$r = +1.0$, yet the daughters are ten points above their fathers in IQ, so the experience of being raised by foster parents had a powerful effect. There is thus no contradiction between the assertion that a trait is perfectly heritable and the assertion that it can be changed radically by environment. As we shall see, this is not a hypothetical example.

Second, a correlation between two variables is not a reliable guide to causation. If A and B are correlated, one may be the cause of the other, they may both be the consequence of a common cause, or they may be entirely accidentally related. The number of cigarettes smoked per day is correlated with the chance of lung cancer because smoking is a cause of lung cancer. The floor area of a person's house and the average age to which he or she lives are positively correlated not because living in a big house is conducive to health but because both characteristics are a consequence of the same cause—high income. For that matter, the distance of the Earth from Halley's comet and the price of fuel are negatively correlated in recent years because one has been decreasing while the other increased, but for totally independent reasons.

In general, heritability is estimated from the correlation of a trait between relatives. Unfortunately, in human populations two important sources of correlation are conflated: Relatives resemble each other not only because they share genes but also because they share environments. This is a problem that can be circumvented in experimental organisms, where genetically related individuals can be raised in controlled environments, but human families are not rat cages. Parents and their offspring may be more similar than unrelated persons because they share genes but also because they share family environment, social class, education, language, etc. To solve this problem, human geneticists and psychologists have taken advantage of special circumstances that are meant to break the tie between genetic and environmental similarity in families.

The first circumstance is adoption. Are particular traits in adopted children correlated with their biological families even when they have been separated from them? Are identical (i.e. monozygotic, or one-egg) twins, separated at birth, similar to each other in some trait? If so, genetic influence is implicated. The second circumstance holds environment constant but changes genetic relationship. Are identical twins more alike than fraternal (i.e. dizygotic, or two-egg) twins? Are two biological brothers or sisters (sibs) in a family more alike than two adopted children in a family? If so, genes are again implicated because, in theory, identical twins and fraternal twins have equal environmental similarity but they are not equally related genetically.

The difficulty with both these kinds of observations is that they only work if the underlying assumptions about environment are true. For the adoption studies to work, it must be true that there is no correlation between the adopting families and the biological families. There must not be selective placement of adoptees. In the case of one-egg and two-egg twins, it must be true that identical twins do not experience a more similar environment than fraternal twins. As we shall see, these problems have been largely ignored in the rush to demonstrate the heritability of IQ.

The theory of estimating heritability is very well worked out. It is well known how large samples should be to get reliable estimates. The designs of the observations to avoid selective adoptions, to get objective measures of test performance without bias on the part of the investigator, to avoid statistical artifacts that may arise from unrepresentative samples of adopting families, are all well laid out in textbooks of statistics and quantitative genetics. Indeed, these theories are constantly put into practice by animal breeders who would be unable to have their research reports published in genetics journals unless they adhered strictly to the standard methodological requirements. The record of psychometric observations on the heritability of IQ is in remarkable contrast. Inadequate sample sizes, biased subjective judgments, selective adoption, failure to separate so-called "separated twins," unrepresentative samples of adoptees, and gratuitous and untested assumptions about similarity of environments are all standard characteristics in the literature of IQ genetics. There has even been, as we shall see, massive and influential fraud. We will review in some detail the state of psychometric genetic observations—not simply because it calls into question the actual heritability of IQ, but because it raises the far more important issue of why the canons of scientific demonstration and credibility should be so radically different in human genetics than in the genetics of pigs. Nothing demonstrates more clearly how scientific methodology and conclusions are shaped to fit ideological ends than the sorry story of the heritability of IQ.

The Cyril Burt Scandal

The clearest evidence, by far, for the genetic determination of IQ was the massive life's work of the late Sir Cyril Burt. In 1969 Arthur Jensen quite correctly referred to Burt's work as "the most satisfactory attempt" to estimate the heritability of IQ. When Burt died, Jensen referred to him as "a born nobleman," whose "larger, more representative samples than any other investigator in the field has ever assembled" would secure his "place in the history of science."[13] Hans Eysenck wrote that he drew "rather heavily" on Burt's work, citing "the outstanding quality of the design and the statistical treatment in his studies."[14]

The Burt data seemed so impressive for a number of very good reasons. First, one of the simplest ways, at least in theory, of demonstrating the heritable basis of a trait is to study separated identical twins. The separated twin pairs have identical genes, and they are assumed not to have shared any common environment. Thus, if they resemble one another markedly in some respect, the resemblance must be due to the only thing they share in common: their identical genes. The largest IQ study of separated identical twins ever reported, supposedly based on fifty-three twin pairs, was that of Cyril Burt. The IQ correlation of separated twin pairs reported by Burt was strikingly high, more so than that reported in the three other studies of separated twins. The most important aspect of Burt's study, however, was that he alone had been able to measure quantitatively the similarity of the environments in which the separated twin pairs had been reared. The incredible (and convenient) result reported by Burt was that there was no correlation at all between the environments of the separated pairs.

Further, in order to fit a genetic model to IQ data, it is necessary to know what the IQ correlations are for a considerable number of types of relatives—some close and some not so close. Burt was the only investigator in history who claimed to have administered the same IQ test, in the same population, to the full gamut of biological relatives of all degrees of closeness. In fact, for some types of relatives (grandparent-grandchild, uncle-nephew, second cousin pairs), the IQ correlations reported by Burt are the *only* such correlations *ever* to have been reported. The Burt correlations for all types of relatives corresponded with remarkable precision to the values expected if IQ were almost entirely determined by the genes.

The blunt fact is that Burt's data, which have played so important a role, were reported and published in what is clearly a truly scandalous and suspicious fashion. The implausibility of Burt's claims should have been noted at once by any reasonably alert and conscientious scientific reader. To begin with, Burt never provided even the most elementary description of how, when, or where his "data" had been collected. The normal canons of scientific reporting were ignored entirely by Burt, and by the editors of the journals that published his papers. He never even identified the "IQ test" he supposedly administered to untold thousands of pairs of relatives. Within many of his papers, even the sizes of his supposed samples of relatives were not reported. The correlations were given without any supporting details. The 1943 paper that first reported many of the correlations between relatives made only the following reference to procedural details: "Some of the inquiries have been published in LCC reports or elsewhere; but the majority remain buried in typed memoranda or degree theses."[15] Conscientious scientists usually do not refer interested readers to their primary sources and documentation in such a cavalier way. The reader should not be surprised by the fact that none of the London County Council reports, typed memoranda, or degree theses glancingly referred to by Burt have ever come to light.

The very few occasions when Burt made specific statements about his procedure should have provoked some doubts in his scientific readers. For example, in a 1955 paper Burt described the procedure by which he obtained IQ test results for parent-child, grandparent-grandchild, uncle-nephew, etc. The IQ data for children were supposedly obtained by revising (on the basis of teachers' comments) the results of unspecified IQ tests given in school. But how did Burt obtain "IQs" for adults? He wrote: "For the assessments of the parents we relied chiefly on personal interviews; but in doubtful or borderline cases an open or a camouflaged test was employed."[16] That is, in measuring the "IQs" of adults Burt did not even *claim* to have administered an objective, standardized IQ test. The IQ was said to have been guessed at during an interview! The spectacle of Professor Burt administering "camouflaged" IQ tests while chatting with London grandparents is the stuff of farce, not of science. The correlations reported by Burt on this claimed basis, however, were routinely presented as hard scientific truths in textbooks of psychology, of genetics, and of education. Professor Jensen referred to precisely this work as "the most satisfactory attempt" to estimate the heritability of IQ. When Burt's procedure was publicly criticized, Hans Eysenck was able to write in Burt's defense: "I could only wish that modern workers would follow his example."[17]

The collapse of Burt's claims within the scientific community began when attention was drawn to some numerical impossibilities in Burt's published papers.[18] For example, Burt in 1955 claimed to have studied 21 pairs of separated identical twins and reported that, on some unnamed group test of intelligence, their IQ correlation was .771. By 1958 the number of pairs had been increased to "over 30"; surprisingly, the IQ correlation remained precisely .771. By 1966, when the sample size had been increased to 53 pairs, the correlation was still exactly .771! This remarkable tendency for IQ correlations to remain identical to the third decimal place was also true of Burt's studies of nonseparated identical twin pairs; as the sample size increased progressively with time, the correlation failed to change. The same identity to the third decimal place was also true of IQ correlations for other types of relatives published by Burt, as sample sizes increased (or in some cases decreased) over time. These and other characteristics indicated that, at the very least, Burt's data and claimed results could not be taken seriously. As one of us in 1974 concluded after surveying Burt's work: "The numbers left behind by Professor Burt are simply not worthy of our current scientific attention."[19]

The scientific exposure of Burt prompted Professor Jensen to execute a brisk about-face. Two years earlier Jensen had described Burt as a born nobleman, whose large and representative samples had secured his place in the history of science. But in 1974 Jensen wrote, after citing the absurdities that critics had already documented, that Burt's correlations were "useless for hypothesis testing"—that is to say, worthless.[20] But Jensen maintained that Burt's work had merely been careless, not fraudulent; and he also maintained that the elimination of Burt's data did not substantially reduce the weight of the evidence demonstrating a high heritability of IQ. That incredible claim was made despite Jensen's earlier assertion that Burt's was "the most satisfactory attempt" to calculate the heritability of IQ.[21]

The argument over Burt's data might have remained a discreet academic affair and might have tiptoed around the question of Burt's fraudulence were it not for the medical correspondent of the London *Sunday Times*, Oliver Gillie. Gillie tried to locate two of Burt's research associates, the Misses Conway and Howard, who had supposedly published papers in a psychological journal edited by Burt. According to Burt, they were responsible for the IQ testing of the separated identical twins, for the testing of other types of relatives, and for much of Burt's published data analyses. But Gillie could uncover absolutely no documentary record of the existence of these research associates. They had not been seen by, and were wholly unknown to, Burt's closest co-workers. When asked about them by his housekeeper, Burt had replied that they had emigrated to Australia or New Zealand, this at a time *before*, according to Burt's published papers, they were testing twins in England. Burt's secretary indicated that Burt had sometimes written papers signed by either Conway or Howard. These facts led Gillie to suggest, in a front-page article in 1976, that Conway and Howard may never have existed.[22] The article flatly accused Burt of perpetrating a major scientific fraud, a charge subsequently supported by two of Burt's former students, now themselves prominent psychometricians, Alan and Ann Clarke.

The public exposure of Burt's fraudulence seemed to strike a raw hereditarian nerve. Professor Jensen wrote that the attack on Burt was designed "to wholly discredit the large body of research on the genetics of human mental abilities. The desperate scorched-earth style of criticism we have come to know in this debate has finally gone the limit, with charges of 'fraud' and 'fakery' now that Burt is no longer here to ... take warranted legal action against such unfounded defamation."[23] Professor Eysenck joined in by pointing out that Burt had been "knighted for his services" and that the charges against him contained "a whiff of McCarthyism, of notorious smear campaigns, and of what used to be known as character assassination."[24]

The attempt to defend Burt by assaulting his critics soon collapsed. The eulogy at Burt's memorial service had been delivered by an admirer, Professor Leslie Hearnshaw, and had prompted Burt's sister, in 1971, to commission Hearnshaw to write a biography of her distinguished brother and to make Burt's private papers and diaries freely available to him. When the fraud charges exploded, Hearnshaw wrote to the *Bulletin* of the British Psychological Society, indicating that he would assess all the available evidence and warning that the charges of Burt's critics could not be lightly dismissed. This warning seems to have muted the tone of Burt's more militant hereditarian defenders. Thus, by 1978, Eysenck wrote of Burt: "On at least one occasion he invented, for the purpose of quoting it in one of his articles, a thesis by one of his students never in fact written; at the time I interpreted this as a sign of forgetfulness."[25]

The Hearnshaw biography, published in 1979, has put to rest any lingering doubts about Burt's wholesale faking.[26] The painstaking searches and inquiries made by Hearnshaw failed to unearth any substantial traces of Miss Conway, or Miss Howard, or of any separated twins. There were many instances of dishonesty, of evasion, and of contradiction in Burt's written replies to correspondents who had inquired about his data. The evidence made clear that Burt had collected no data at all during the last thirty years of his life, when, supposedly, most of the separated twins had been studied. With painful reluctance, Hearnshaw found himself forced to conclude that the charges made by Burt's critics were "in their essentials valid." The evidence demonstrated that Burt had "fabricated figures" and had "falsified." There is now no doubt whatever that all of Burt's "data" on the heritability of IQ must be discarded. The loss of these incredibly clear-cut "data" has been devastating to the claim that a substantial IQ heritability was demonstrated.

But what are we to make of the additional fact that Burt's transparently fraudulent data were accepted for so long, and so uncritically, by the "experts" in the field? Perhaps the clearest moral to be drawn from the Burt affair was spelled out by N. J. Mackintosh in his review of the Hearnshaw biography in the *British Journal of Psychology*:

Ignoring the question of fraud, the fact of the matter is that the crucial evidence that his data on IQ are scientifically unacceptable does not depend on any examination of Burt's diaries or correspondence. It is to be found in the data themselves. The evidence was there . . . in 1961. It was, indeed, clear to anyone with eyes to see in 1958. But it was not seen until 1972, when Kamin first pointed to Burt's totally inadequate reporting of his data and to the impossible consistencies in his correlation coefficients. Until then the data were cited, with respect bordering on reverence, as the most telling proof of the heritability of IQ. It is a sorry comment on the wider scientific community that "numbers . . . simply not worthy of our current scientific attention" . . . should have entered nearly every psychological textbook.[27]

We do not view the uncritical acceptance of Burt's data as an unusual or inexplicable "sorry comment on the wider scientific community." The fraud perpetrated by Burt, and unwittingly propagated by the scientific community, served important social purposes. Professor Hearnshaw's biography essentially saves the face of psychometry by probing the individual psychology of Burt to ask why he should have been moved to such fraudulence. Burt, no longer a nobleman but now victim of a debilitating and psychiatrically distressing disorder, has become the bad apple of psychometry. By 1980, when the British Psychological Society was prepared to draw up its "Balance Sheet on Burt,"[28] there had been a closing of the ranks; the psychometric doyens reiterated their belief that, despite the eviction of Burt, the residual evidence for the heritability of intelligence was strong. The social function of IQ ideology was still dominant.

Separated Identical Twins

With Burt out of the way, there have in fact been three reported studies of the IQs of separated identical twins. The largest study, by Shields in England, reported an IQ correlation of .77.[29] The American study by Newman, Freeman, and Holzinger found a correlation of .67,[30] while a small-scale Danish study by Juel-Nielsen reported a correlation of .62.[31] Taken at face value, these studies would suggest a substantial heritability of IQ. There are many reasons, however, why they should not be taken at face value.

To begin with, it is obvious that the sample of "separated" identical twins studied by psychologists must be highly biased. There presumably exist some pairs of identical twins who have been separated at birth and who do not know of one another's existence. These genuinely separated twins cannot, of course, respond to the appeals of scientists for separated twins to volunteer to be studied. The Shields study, for example, located its subjects by use of a television appeal. The "separated" twins located in this way in fact included 27 pairs in which the two twins had been reared in related branches of the same biological family. There were only 13 pairs in which the two twins had been reared in unrelated families. The most common pattern was for the biological mother to rear one of the twins, with the other twin being reared by the maternal grandmother or by an aunt.

From the raw data it can be calculated that the IQ correlation of the 27 pairs reared in the same family network was .83, significantly higher than the correlation of .51 in the 13 pairs reared in unrelated families. This significant difference is obviously an environmental effect; recall that each twin pair was genetically identical. The data make clear that genetically identical twins reared in the same family network, and thus sharing similar environmental experiences, are much more alike than genetically identical twins reared in unrelated families. Further, it should not be supposed that the correlation of .51 observed among twins reared in unrelated families is unambiguous evidence for some heritability of IQ. The most common pattern, even among pairs reared in unrelated families, was for the mother to raise one twin while the other twin was raised by close family friends. There is thus no reason to assume that any of the Shields twins were reared in very different social conditions. We have no way of knowing what the IQ correlation would be in a set of identical twins who had been separated at birth and randomly placed in two families randomly chosen from the full range of rearing environments provided by English society, but we can deduce that the correlation found in such a science-fiction experiment would be considerably less than .51, and it might in fact be zero.

The reader whose knowledge of separated twin studies comes only from the secondary accounts provided by textbooks can have little idea of what, in the eyes of the original investigators, constitutes a pair of "separated" twins. To be included in the Shields study, for example, it was only necessary that the two twins, at some time during childhood, had been reared in different homes for at least five years. The following examples, taken from Shields's case histories, are illuminating.

Jessie and Winifred had been separated at three months. "Brought up within a few hundred yards of one another, . . . told they were twins after the girls discovered it for themselves, having gravitated to one another at school at the age of five. . . . They play together quite a lot. . . . Jessie often goes to tea with Winifred. . . . They were never apart, wanted to sit at the same desk. . . ." Ironically, the investigator who has supplied us with more than half the documented cases of "separated" twins here informs us that a separated pair of 8-year-olds "were never apart." The technical use of the word "separated" by the scientists of IQ obviously differs from the usage of the same word by ordinary people. We might note, also, that Jessie and Winifred had been reared by unrelated families. Presumably a twin pair reared by related families would be even less separated.

Bertram and Christopher had been separated at birth. "The paternal aunts decided to take one twin each and they brought them up amicably, living next door to one another in the same Midlands colliery village. . . . They are constantly in and out of each other's houses." Odette and Fanny, on the other hand, had been separated only between the ages of three and eight. During that period they changed places every six months, one going to the mother, the other to the maternal grandmother. Benjamin and Ronald had been "brought up in the same fruit-growing village, Ben by his parents, Ron by the grandmother. . . . They were at school together. . . . They have continued to live in the same village." The twins were fifty-two years old when they traveled to London to be IQ tested by Shields. Finally, consider Joanna and Isabel, aged fifty, who had been "separated from birth to five years" but who then "went to private school together."

The study of separated identical twins would be of theoretical value if it could be assumed that there was little or no systematic similarity between the environments in which pair members had been reared. Professor Burt, without providing any of the details, was indeed able to report that there was *no* correlation between the environments of his mythical separated pairs. The real-life case studies provided by Shields, however, make clear that in the actual world the environments of so-called separated twins have been massively correlated. That fact alone makes such studies virtually worthless for attempts to demonstrate the heritability of IQ.

The fatal flaw of highly correlated environments is obvious in all three of the studies of separated twins. Thus, in the American study of nineteen twin pairs by Newman et al., Kenneth and Jerry had been adopted by two different families. Kenneth's foster father was "a city fireman with a very limited education." Jerry's foster father, by contrast, was "a city fireman with only fourth-grade education." The two boys had lived between the ages of five and seven in the same city in which their fathers worked but were said to be "unaware of the fact." Harold and Holden, another pair studied by Newman et al., had each been adopted by a family relative. They lived three miles apart and attended the same school.

The Juel-Nielsen study of twelve Danish pairs included Ingegard and Monika, each cared for by relatives until the age of seven. They then lived together with their mother until they were fourteen. "They were usually dressed alike and very often confused by strangers, at school, and sometimes also by their step-father. . . . The twins always kept together when children, they played only with each other and were treated as a unit by their environment." Remember that these and similar separated twin pairs are the bedrock upon which the scientific study of the heritability of IQ has been based. The ludicrous shortcomings of these studies are obvious to the most naive of nonscientific eyes. Perhaps only a scientist caught up with an enthusiasm for an abstract idea and trained to accept the "objectivity" of numbers could take such studies seriously.

There are other severe problems with the separated twin studies, which have been documented in full elsewhere.[32] For example, in each study the usual procedure has been for the same investigator to administer the IQ test to both members of a twin pair. This violates the basic methodological requirement that such testing should be done "blind." That is, Twin B should be tested by a person who has no knowledge of Twin A's IQ score; otherwise the administration and/or scoring of the test to Twin B may be biased by the tester's knowledge of Twin A's score. There is, in fact, suggestive evidence that such unconscious tester bias, a very common finding in research involving human subjects, has inflated the correlations reported in twin studies. Finally, we should note that the investigators in these studies have depended heavily on the verbal accounts of the volunteer twins themselves to provide details about the conditions and duration of their separation. There is evidence that the twins sometimes tend romantically to exaggerate the degree of their separation, and "facts" reported by the twins have sometimes been mutually contradictory. When all these problems are added to the overwhelming flaw of highly correlated environments, and when it is recalled that the apparently most impressive study has been unmasked as a fraud, it seems clear that the study of separated identical twins has failed to demonstrate a heritable basis for IQ test scores.

Studies of Adopted Children

The fact that in ordinary families parents and children resemble one another in IQ does not in itself say anything about the relative importance of heredity and environment. As should be clear by now, the problem is that the parent provides the child both with its genes and with its environment. The high-IQ parent, who has transmitted his or her genes to a child, is also likely to provide that child with intellectual stimulation in the home and to stress the importance of doing good schoolwork. The practice of adoption makes possible, at least in theory, a separation of genetic from environmental transmission. The adoptive parent provides his or her child with an environment, while the genes, of course, come from the child's biological parents. Thus, the IQ correlation between adopted child and adoptive parent has been of particular interest to investigators of IQ heritability, especially when it is compared to other relevant IQ correlations. The key question, as we shall see, is: To what other correlations can the correlation between adoptive parent and adoptive child be meaningfully compared?

Two early and influential studies of adoption by Burks[33] and Leahy[34] employed identical experimental designs. This "classical" design is schematically illustrated in Figure 5.1. First, Burks and Leahy calculated the IQ correlation, in a set of adoptive families, between adoptive parents and adopted children. The correlation, taken to reflect the effects of environment alone, averaged out to a mere .15. That correlation was then compared to the correlation between biological parent and biological child observed in a "matched control group" of ordinary families. The latter correlation, presumed to reflect the effects of environment *plus* genes, averaged out to a full .48. The comparison between the two correlations was said to demonstrate that, although environment plays some small role, heredity is far more important as a determiner of IQ.

This comparison makes sense, however, only if we are willing to believe that the biological families used as control groups in these

FIGURE 5.1 / The "classical" adoption design of Burks and Leahy. Note that correlations in two different, but supposedly matched, groups of families are compared. In the biological families, parent transmits environment *plus* genes to child.

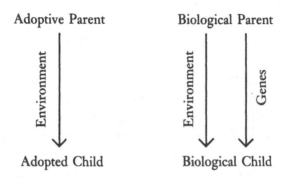

studies were in fact meaningfully "matched" to the adoptive families. There are some obvious ways in which adoptive families must, as a group, differ from ordinary biological families. For one thing, all the adoptive parents, but not necessarily all the biological parents, actively wanted children. For another, adoptive parents, by law, are carefully screened by adoption agencies before they are allowed to adopt and therefore as a group tend to be especially suitable parents, although there are, of course, exceptions. They are selected as being emotionally stable, economically secure, not alcoholics, without criminal records, etc. Thus adoptive families generally provide much better than average environments for their children; as well, adoptive parents often have quite high IQ scores as a consequence of their own childhood advantages. The key fact for present purposes is that there will be very little variation in the richness of the environments provided by adoptive parents. The necessary statistical consequence of this is that there cannot be a very high correlation between adopted children's IQ and any environmental measure, such as the adoptive parents' IQ. Where environment does not vary, or varies very little, it cannot be systematically correlated with the child's IQ. The "matched" control groups of biological families, who have not been rigorously selected by adoption agencies, will doubtless exhibit more variation in the environments they provide for their children. That, of course, allows for a higher parent-child correlation in the biological families.

To be sure, Burks and Leahy each attempted to match their adoptive and biological families in at least some ways. The two groups of children had been matched for age and sex. The two types of families had been matched for parental occupation, for parental educational levels, and for "type of neighborhood." The adoptive parents, however, were considerably older than the control parents; they had tried to have their own biological child for some time before adopting. For obvious reasons, there were significantly fewer siblings in the adoptive than in the biological families. The income of the adoptive families turns out to have been 50 percent higher. The homes of the adoptive parents, with smaller families, were larger and 50 percent more expensive than those of the "matched" biological parents. Thus, despite apparently careful matching, these differences doubtless reflect the fact that adoptive parents as a group are relatively "successful" people. They make clear that adoptive and biological families cannot meaningfully be regarded as "matched" merely because they are comparable on a few rough demographic measures. There is clear evidence in the Burks and Leahy studies that the environments of the adoptive families were not only richer but also much less variable than the environments of the biological families.[35] These considerations mean that a comparison of correlations across the adoptive and biological families has no theoretical point.

There is, however, an obvious possible improvement on the "classical" design of Burks and Leahy, illustrated schematically in Figure 5.2, which avoids the impossible requirement of matching adoptive and biological families. There are many adoptive parents who, in addition to adopting a child, also have a biological child of their own. Thus, in a sample of such families, it is possible to correlate a parent's IQ with the IQ of (a) its adopted child and (b) its biological child. The two children, in such a comparison, have been reared in the same household by the same parents. To the extent that genes determine IQ, the correlation between parent and biological child should obviously be larger than that between parent and adoptive child. The parents in all such families have been carefully selected by adoption agencies; we therefore expect relatively little environmental variation and relatively small IQ correlations between parent and child in such a study. The virtue of the new design, however, is that this should be true for *both* the adoptive and the biological correlations studied in the same group

FIGURE 5.2 / The new adoption design of Scarr and Weinberg (1977) and of Horn et al., (1979). Note that only one set of families is involved, with each family containing both an adopted and a biological child. The parent transmits environment *plus* genes to the biological child.

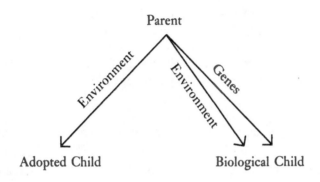

of families. There is plenty of room for any genetic effect to display itself in a higher correlation for the biological parent-child pairs.

Two recent studies have employed the new design: one in Minnesota in 1977 by Scarr and Weinberg[36] and one in Texas in 1979 by Horn, Loehlin, and Willerman.[37] The investigators in each case were behavior geneticists who clearly expected to discover evidence supporting a high heritability of IQ.

The results for mother-child pairings in both studies are as follows: The *same* mother's IQ, remember, has been correlated with the IQ of her adopted and of her biological child. There is no significant difference between the two correlations: In Texas the mother was a trifle more highly correlated with her adopted child, and in Minnesota with her biological child. The Minnesota study, it might be noted, was based upon transracial adoptions. That is, in almost all cases the mother and her biological child were both white, while her adopted child was black. The child's race, like its adoptive status, had no effect on the degree of parent-child resemblance in IQ. These results appear to inflict fatal damage to the notion that IQ is highly heritable. Children reared by the same mother resemble her in IQ to the same degree, whether or not they share her genes.

The results for father-child pairings are not so clear-cut. Though not statistically significant, they are more easily compatible with the notion that IQ may be partly heritable. However, when we turn to the IQ

TABLE 5.1 / Mother-child IQ correlations in adoptive families containing biological children.

	Texas Study	Minnesota Study
Mother × Biological Child	0.20 (N = 162)	0.34 (N = 100)
Mother × Adopted Child	0.22 (N = 151)	0.29 (N = 66)

"N" refers to the number of mother-child pairings on which each tabled correlation is based. Texas study is Horn et al. (37); Minnesota study is Scarr and Weinberg (36).

correlations between the various types of siblings found in these families, they are again entirely inconsistent with the notion that IQ is significantly heritable. In these families there are some pairs of biological related siblings (the biological children of the adoptive parents); there are also genetically unrelated pairs of adoptive siblings (two children adopted by the same parents); finally, there are genetically unrelated pairs made up of a biological and an adoptive child of the same parents. The correlations for all sibling types show no differences.

MZ Twins, DZ Twins,
and Other Kinships

By far the most common type of heritability study involves comparing the two fundamentally different types of twins, monozygotic (MZ) and dizygotic (DZ). Remember that the MZ twins result from the fertilization of a single ovum by a single sperm. There is an extra split of the zygote early in development, resulting in the birth of two genetically identical individuals, always of the same sex and typically, but not always, strikingly similar in appearance. The DZ twins occur when two separate sperm fertilize two separate ova at about the same time. The mother gives birth to two individuals, but the two are no more alike genetically than are ordinary siblings. The DZ twins, like ordinary siblings, share on average about 50 percent of their genes.

They may be either of the same or of different sexes, and their physical resemblance is no greater than that of ordinary siblings.

The fact that MZ twins are twice as similar genetically than are DZ twins leads us to expect that, for any genetically determined trait, the correlation between pairs of MZ twins should be greater than that between pairs of same-sexed DZ twins. (We restrict the comparison to same-sexed DZs since all MZs are same-sexed, and sex might affect the trait in question.) The degree of heritability of a trait can in theory be estimated from the magnitude of the difference between the MZ and DZ correlations. With a very highly heritable trait, the MZ correlation should approach 1.00, while the DZ correlation approaches .50. Put simply, MZ twins should resemble one another in heritable traits much more than do DZ twins. There have been many dozens of studies comparing the IQ correlations of MZ and DZ twins. With almost no exceptions, the studies demonstrate that the IQ correlation of MZs is considerably higher than that of DZs. Typically, correlations reported for MZs range between .70 and .90, compared to a range of .50 to .70 for same-sexed DZs.

Though hereditarians attribute this difference to the greater genetic similarity of MZs, there are also some obvious environmental reasons to expect higher correlations among MZ than among DZ twins, especially when one realizes the degree to which an MZ pair creates or attracts a far more similar environment than that experienced by other people. Because of their striking physical similarity, parents, teachers, and friends tend to treat them very much alike and often even confuse them for one another. MZ twins tend to spend a great deal of time with one another, doing similar things, much more so than is the case with same-sexed DZ twins, as established by many questionnaire studies. The MZ twins are much less likely to have spent a night apart from each other during childhood. The MZ twins are more likely to dress similarly, to play together, and to have the same friends. When Smith questioned twins, 40 percent of MZs reported that they usually studied together, compared to only 15 percent of DZs.[38] In an extreme example of this deliberate pattern, one of the most extraordinary social experiences of identical twins is the institution of the twin convention, to which identical twins of all ages go, or are sent by their parents, dressed identically, acting identically, to show off their identity, and, in a sense, to compete with other twins to see who can be the most "identical."

There is no great imagination required to see how such a difference between MZs and DZs might produce the reported difference in IQ correlations. It is entirely clear that the environmental experiences of MZs are much more similar than those of DZs.

Twin studies as a whole, then, cannot be taken as evidence for the heritability of IQ. They have been interpreted, of course, as if their proof were adequate, and hereditarian scholars have routinely ground out quantitative estimates of IQ heritability from the results of twin studies. Claiming validity for such calculations can only be done by willfully ignoring the obvious fact that MZ and DZ twins differ in environment as well as in genetic similarity.

Heritability and Changeability

The careful examination of the studies of heritability of IQ can leave us with only one conclusion: we do not know what the heritability of IQ really is. The data simply do not allow us to calculate a reasonable estimate of genetic variation for IQ in any population. For all we know, the heritability may be zero or 50 percent. *In fact, despite the massive devotion of research effort to studying it, the question of heritability of IQ is irrelevant to the matters at issue.* The great importance attached by determinists to the demonstration of heritability is a consequence of their erroneous belief that heritability means unchangeability. An American court recently ruled that an advertised cure for baldness was fraudulent on the face of it because baldness is hereditary. But this is simply wrong. The heritability of a trait only gives information about how much genetic and environmental variation exists in the population *in the current set of environments.* It has absolutely no predictive power for the result of changing the set of environments. Wilson's disease, a defect of copper metabolism, is inherited as a single gene disorder and is fatal in early adulthood. It is curable, however, by the administration of the drug penicillamine. IQ variation could be 100 percent heritable in some population, yet a cultural shift could change everyone's performance on IQ tests. In fact, this is what happens in adoption studies: Even when adopted children are not correlated, parent by parent, with their adoptive parents, their IQ scores *as a group* resemble the adoptive parents *as a group* much more than they resemble their biological parents. So, in an adoption study by Skodak and Skeels the mean IQ of the adopted children was 117 while the mean IQ of their biological mothers was only 86.[39] A similar result was reported in a study of children in English residential nursery homes.[40] Children who remained in the homes had an average IQ of 107, those adopted out of the homes an IQ of 116, but those returned to their biological mothers, only 101. The most striking and consistent observation in adoption studies is the raising of IQ, irrespective of any correlation with adoptive or biological parents. The point is that adoptive parents are not a random sample of households but tend to be older, richer, and more anxious to have children; and, of course, they have fewer children than the population at large. So the children they adopt receive the benefits of greater wealth, stability, and attention. It shows in the childrens' test performances, which clearly do not measure something intrinsic and unchangeable.

The confusion of "heritable" with "unchangeable" is part of a general misconception about genes and development. The phenotype of an organism is changing and developing at all times. Some changes are irreversible and some reversible, but these categories cross those of the heritable and nonheritable. The loss of an eye, an arm, or a leg is irreversible but not heritable. The appearance of Wilson's disease is heritable but not irreversible. The morphological defect that causes blue babies is congenital, nonheritable, irreversible under normal developmental conditions, but reversible surgically. The extent to which morphological, physiological, and mental characteristics do or do not change in the course of individual lifetimes and the history of the species is a matter of historical contingency itself. The variation from person to person in the ability to do arithmetic, whatever its source, is trivial compared to the immense increase in calculating power that has been put into the hands of even the poorest student of mathematics by the pocket electronic calculator. The best studies in the world of the heritability of arithmetic skill could not have predicted that historical change.

The final error of the biological determinists' view of mental ability is to suppose that the heritability of IQ within populations somehow explains the differences in test scores between races and classes. It is claimed that if black and working-class children do worse on an average on IQ tests than white and middle-class children and if the differences are greater than can be accounted for by environmental factors, the differences must be genetically caused. This is the argument of Arthur Jensen's *Educability and Group Differences* and Eysenck's *The Inequality of Man*. What it ignores, of course, is that the causes of the differences between groups on tests are not, in general, the same as the sources of variation within them. There is, in fact, no valid way to reason from one to the other.

A simple hypothetical but realistic example shows how the heritability of a trait within a population is unconnected to the causes of differences between populations. Suppose one takes from a sack of open-pollinated corn two handfuls of seed. There will be a good deal of genetic variation between seeds in each handful, but the seeds in one's left hand are on the average no different from those in one's right. One handful of seeds is planted in washed sand with an artificial plant growth solution added to it. The other handful is planted in a similar bed, but with half the necessary nitrogen left out. When the seeds have germinated and grown, the seedlings in each plot are measured, and it is found that there is some variation in height of seedling from plant to plant within each plot. This variation within plots is entirely genetic because the environment was carefully controlled to be identical for all the seeds within each plot. The variation in height is then 100 percent heritable. But if we compare the two plots, we will find that all the seedlings in the second are much smaller than those in the first. This difference is not at all genetic but is a consequence of the difference in nitrogen level. So the heritability of a trait within populations can be 100 percent, but the cause of the difference between populations can be entirely environmental.

It is an undoubted fact that in the school population at large the IQ performance of blacks and whites differs on the average. Black children in the United States have a mean IQ score of about 85 as compared with 100 for the white population, on which the test was standardized. Similarly, there is a difference in IQ on the average between social classes. The most extensive report on the relation between occupational class and IQ is that of Cyril Burt, so it cannot be used, but other studies have found that the children of professional and managerial fathers score about 15 points higher on the average than children of unskilled laborers. Not uncharacteristically, Burt reported rather larger differences. Is there any evidence that these race and class differences are in part a consequence of genetic differences between groups?

What Is Race?

Before we can sensibly evaluate claims of genetic differences in IQ performance between races, we need to look at the very concept of race itself: What is really known about genetic differences between what are conventionally thought of as human races?

Until the mid-nineteenth century, "race" was a fuzzy concept that included a number of kinds of relationships. Sometimes it meant the whole species, as "the human race"; sometimes a nation or tribe, as "the race of Englishmen"; and sometimes merely a family, as "He is the last of his race." About all that held these notions together was that members of a "race" were related by ties of kinship and that their shared characteristics were somehow passed from generation to generation. With the rise to popularity of Darwin's theory of evolution, biologists soon began to use the concept of "race" in a quite different but no more ultimately consistent way. It simply came to mean "kind," an identifiably different form of organism within a species. So there were light-bellied and dark-bellied "races" of mice, or banded- or unbanded-shell "races" of snails. But defining "races" simply as observable kinds produced two curious contradictions. First, members of different "races" often existed side by side within a population. There might be twenty-five different "races" of beetles, all members of the same species, living side by side in the same local population. Second, brothers and sisters might be members of two different races, since the characters that differentiated races were sometimes influenced by alternative forms of a single gene. So a female mouse of the light-bellied "race" could produce offspring of both light-bellied and dark-bellied races, depending on her mate. Obviously there was no limit to the number of "races" that could be described within a species, depending on the whim of the observer.

Around 1940, biologists, under the influence of discoveries in population genetics, made a major change in their understanding of race. Experiments on the genetics of organisms taken from natural populations made it clear that there was a great deal of genetic variation between individuals even in the same family, not to speak of the same population. Many of the "races" of animals previously described and named were simply alternative hereditary forms that could appear within a family. Different local geographic populations did not differ from each other absolutely, but only in the relative frequency of different characters. So, in human blood groups, some individuals were type A, some type B, some AB, and some O. No population was exclusively of one blood type. The difference between African, Asian, and European populations was only in the proportion of the four kinds. These findings led to the concept of "geographical race" as a population of varying individuals, freely mating among each other but different in average proportions of various genes from other populations. Any local random breeding population that was even slightly different in the proportion of different gene forms from other populations was a geographical race.

This new view of race had two powerful effects. First, no individual could be regarded as a "typical" member of a race. Textbooks of anthropology would often show photographs of "typical" Australian aborigines, tropical Africans, Japanese, etc., listing as many as fifty or a hundred "races," each with its typical example. Once it was recognized that every population was highly variable and differed largely in average proportions of different forms from other populations, the concept of the "type specimen" became meaningless. The second consequence of the new view of race was that since every population differs slightly from every other one on the average, all local interbreeding populations are "races," so race really loses its significance as a concept. The Kikuyu of East Africa differ from the Japanese in gene frequencies, but they also differ from their neighbors, the Masai, and, although the extent of the differences might be less in one case than in the other, it is only a matter of degree. This means that the *social* and *historical* definitions of race that put the two East African tribes in the same "race" but put the Japanese in a different "race" were biologically arbitrary. How much difference in the frequencies of A, B, AB, and O blood groups does one require before deciding it is large enough to declare two local populations are in separate "races"?

The change in point of view among biologists had an eventual effect on anthropology in that about 40 years ago textbooks began to play down the whole issue of defining races, but the changes in academic views have had little effect on everyday consciousness of race. We still speak casually of Africans as one race, Europeans as another, Asians as another, using distinctions that correspond to our everyday impressions. No one would mistake a Masai for a Japanese or either for a Finn. Despite variation from individual to individual within these groups, the differences between groups in skin color, hair form, and some facial features make them clearly different. What racists do is to take these evident differences and claim that they demonstrate major genetic separation between "races." Is there any truth in this assertion? Are the differences in skin color and hair form that we use to distinguish races in our everyday experience typical of the genetic differentiation between groups, or are they for some reason unusual?

We must remember that we are conditioned to observe precisely those features and that our ability to distinguish individuals as opposed to types is an artifact of our upbringing. We have no difficulty at all in telling individuals apart in our own group, but "they" all look alike. The question is, if we could look at a random sample of different genes, not biased by our socialization, how much difference would there be between major geographical groups, say between Africans and Australian aborigines, as opposed to the differences between individuals within these groups? It is, in fact, possible to answer that question.

During the last forty years, using the techniques of immunology and of protein chemistry, geneticists have identified a large number of human genes that code for specific enzymes and other proteins. Very large numbers of individuals from all over the world have been tested to determine their genetic constitution with respect to such proteins, since only a small sample of blood is needed to make these determinations. About 150 different genetically coded proteins have been examined, and the results are very illuminating for our understanding of human genetic variation.

It turns out that 75 percent of the different kinds of proteins are identical in all individuals tested, regardless of population, with the exception of an occasional rare mutation. These so-called *monomorphic* proteins are common to all human beings of all races; the species is essentially uniform with respect to the genes that code them. The other 25 percent, however, are *polymorphic* proteins. That is, there exist two or more alternative forms of the protein, coded by alternative forms of a gene, that are common but at varying frequencies in our species. We can use these polymorphic genes to ask how much difference there is between populations, as compared with the difference between individuals within populations.

An example of a highly polymorphic gene is the one that determines the ABO blood type. There are three alternative forms of the gene, which we will symbolize by A, B, and O, and every population in the world is characterized by some particular mixture proportions of the three. For example, Belgians have about 26 percent A and 6 percent B; the remaining 68 percent is O. Among Pygmies of the Congo, the proportions are 23 percent A, 22 percent B, and 55 percent O. The frequencies can be depicted as a triangular diagram, as shown in Figure 5.3. Each point represents a population, and the proportion of each gene form can be read as the perpendicular distance from the point to the appropriate side of the triangle. As the figure shows, all human populations are clustered fairly close together in one part of the frequency space. There are no populations, for example, with very high B and very low A and O (lower right-hand corner). The figure also shows that populations that belong to what we call major "races" in our everyday usage do not cluster together. The dashed lines have been put around populations that are similar in ABO frequencies, but these do not mark off racial groups. For example, the cluster made up of populations 2, 8, 10, 13, and 20 include an African, three Asian, and one European population.

A major finding from the study of such polymorphic genes is that none of these genes perfectly discriminates one "racial" group from another. That is, there is no gene known that is 100 percent of one form in one race and 100 percent of a different form in some other race. Reciprocally, some genes that are very variable from individual to individual show no average difference at all between major races. Table 5.2 shows the three polymorphic genes that are most different between "races" and the three that are most similar among the "races." The first column gives the name of the protein or blood group, and the second column gives the symbols of the alternative forms (*alleles*) of the gene that is varying. As the table shows, there are big differences in relative frequencies of the alleles of the Duffy, Rhesus, and P blood groups from "race" to "race," and there may be an allele like Fy^b that is found only in one group, but no group is "pure" for any genes. In contrast,

FIGURE 5.3 / A triallelic diagram of the ABO blood group allele frequencies for human populations. Each point represents a population: the perpendicular distances from the point to the sides represent the allele frequencies as indicated in the small triangle. Populations 1–3 are African, 4–7 are American Indians, 8–13 are Asians, 14–15 are Australian aborigines, and 16–20 are Europeans. Dashed lines enclose arbitrary classes with similar gene frequencies, which do not correspond to the "racial" classes. (Jacquard, 1970.)

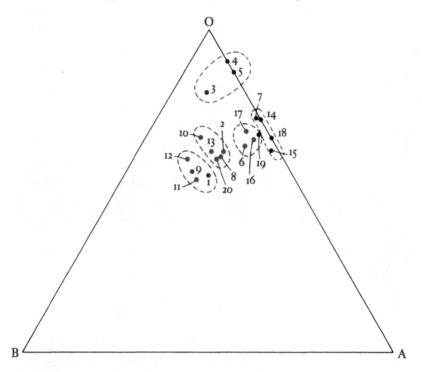

the Auberger, Xg, and Secretor proteins are very polymorphic within each "race," but the differences between groups is very small. It must be remembered that 75 percent of known genes in humans do not vary at all, but are totally monomorphic throughout the species.

Rather than picking out the genes that are the most different or the most similar between groups, what do we see if we pick genes at random? Table 5.3 shows the outcome of such a random sample. Seven enzymes known to be polymorphic were tested in a group of Europeans and Africans (actually black Londoners who had come from West Africa and white Londoners). In this random sample of genes there is a remarkable similarity between groups. With the exception

TABLE 5.2 / Examples of extreme differentiation and close similarity in blood-group allele frequencies in three racial groups

Gene	Allele	POPULATION: Caucasoid	Negroid	Mongoloid
Duffy	Fy	0.0300	0.9393	0.0985
	Fy^a	0.4208	0.0607	0.9015
	Fy^b	0.5492	0.0000	0.0000
Rhesus	R_0	0.0186	0.7395	0.0409
	R_1	0.4036	0.0256	0.7591
	R_2	0.1670	0.0427	0.1951
	r	0.3820	0.1184	0.0049
	r'	0.0049	0.0707	0.0000
	Others	0.0239	0.0021	0.0000
P	P_1	0.5161	0.8911	0.1677
	P_2	0.4839	0.1089	0.8323
Auberger	Au^a	0.6213	0.6419	—
	Au	0.3787	0.3581	—
Xg	Xg^a	0.67	0.55	0.54
	Xg	0.33	0.45	0.46
Secretor	Se	0.5233	0.5727	—
	se	0.4767	0.4273	—

Source: From a summary provided in L. L. Cavalli-Storza and W. F. Bodmer, *The Genetics of Human Populations* (San Francisco: Freeman, 1971), pp. 724–31. See this source for information on other loci and for data sources.

of phosphoglucomutase-3, for which there is a reversal between groups, the most common form of each gene in Africans is the same form as for the Europeans, and the proportions themselves are very close. Such a result would lead us to conclude that the genetic difference between blacks and whites is negligible as compared with the polymorphism within each group.

The kind of question asked in Table 5.3 can in fact be asked in a very general way for large numbers of populations for about twenty genes that have been widely studied all over the world. Suppose we measure

TABLE 5.3 / Allelic frequencies at seven polymorphic loci in Europeans and black Africans

Locus	EUROPEANS: Allele 1	Allele 2	Allele 3	AFRICANS: Allele 1	Allele 2	Allele 3
Red cell acid phosphatase	0.36	0.60	0.04	0.17	0.83	0.00
Phospho-glucomutase 1	0.77	0.23	0.00	0.79	0.21	0.00
Phospho-glucomutase-3	0.74	0.26	0.00	0.37	0.63	0.00
Adenylate kinase	0.95	0.05	0.00	1.00	0.00	0.00
Peptidase A	0.76	0.00	0.24	0.90	0.10	0.00
Peptidase D	0.99	0.01	0.00	0.95	0.03	0.02
Adenosine deaminase	0.94	0.06	0.00	0.97	0.03	0.00

Source: R. C. Lewontin, *The Genetic Basis of Evolutionary Change* (New York: Columbia Univ. Press, 1974). Adapted from H. Harris, *The Principles of Human Biochemical Genetics* (Amsterdam and London: North-Holland, 1970).

the variation among humans for some particular gene by the probability that a gene taken from one individual is a different alternative form (allele) than that taken from another individual at random from the human species as a whole. We can then ask how much less variation there would be if we chose the two individuals from the same "race." The difference between the variation over the whole species and the variation within a "race" would measure the proportion of all human variation that is accounted for by racial differences. In like manner we could ask how much of the variation within a "race" is accounted for by differences between tribes or nations that belong to the same "race," as opposed to the variation between individuals within the same tribe or nation. In this way we can divide the totality of human genetic variation into a portion between individuals within populations, between local populations within major "races," and between major "races." That calculation has been carried out independently by three different groups of geneticists using slightly different data and somewhat different statistical methods but with the identical result. Of all human genetic variation known for enzymes and other proteins, where it has been possible to actually count up the frequencies of different forms of the genes and so get an objective estimate of genetic variation, 85 percent turns out to be between individuals within the same local population, tribe, or nation; a further 8 percent is between tribes or nations within a major "race"; and the remaining 7 percent is between major "races." That means that the genetic variation between one Spaniard and another, or between one Masai and another, is 85 percent of all human genetic variation, while only 15 percent is accounted for by breaking people up into groups. If everyone on earth became extinct except for the Kikuyu of East Africa, about 85 percent of all human variability would still be present in the reconstituted species. A few gene forms would be lost—like the Fy^b allele of the Duffy blood group that is known only in European, or the Diego blood factor known only in American Indians—but little else would be changed.

The reader will have noticed that to carry out the calculation of partitioning variation between "races," some method must have been used for assigning each nation or tribe to a "race." The problem of what one means by a "race" comes out forcibly when making such assignments. Are the Hungarians European? They certainly *look* like Europeans, yet they (like the Finns) speak a language that is totally unrelated to European languages and belongs to the Turkic family of languages from Central Asia. And what about the modern-day Turks? Are they Europeans, or should they be lumped with the Mongoloids? And then there are the Urdu- and Hindi-speaking people of India. They are the descendants of a mixture of Aryan invaders from the north, the Persians from the west, and the Vedic tribes of the Indian subcontinent. One solution is to make them a separate race. Even the Australian aborigines, who have often been put to one side as a separate race, mixed with Papuans and with Polynesian immigrants from the Pacific well before Europeans arrived. No group is more hybrid in its origin than the present-day Europeans, who are a mixture of Huns, Ostrogoths, and Vandals from the east, Arabs from the south, and Indo-Europeans from the Caucasus. In practice, "racial" categories are established that correspond to major skin color groups, and all the borderline cases are distributed among these or made into new races according to the whim of the scientist. But it turns out not to matter much how the groups are assigned, because the differences between major "racial" categories, no matter how defined, turn out to be small. Human "racial" differentiation is, indeed, only skin deep. Any use of racial categories must take its justifications from some other source than biology. The remarkable feature of human evolution and history has been the very small degree of divergence between geographical populations as compared with the genetic variation among individuals.

IQ Differences Between Groups

The only way to answer the question of genetic differences in IQ between groups would be to study adoption across racial and class boundaries. Such studies are not easy to find, but the several that have been done all give the same result. In the study by Tizard[41] of black, white, and mixed-parentage children in English residential nurseries, using three preschool tests of mental performance, the differences were not larger than could be expected from statistical variations due to chance; but, taken at face value, blacks and mixed-parentage children did *better* than whites. Another relevant case is the comparison of the children of black and of white U.S. soldiers and German mothers who were left behind to be raised in Germany when their fathers returned home after the Occupation. Again, there is a small difference favoring the black children. Two studies comparing the amount of white ancestry of black children with their IQ scores found no correlation. On the other hand, a study of black children adopted by white families showed a much higher IQ than for children in the general population, but within these adoptees, children of two black parents performed less well than when one of the biological parents was black and one white.[42] In fact, this is the sum total of evidence on genetic differences between blacks and whites that makes any effort at all to separate the genetic from the social.

Like all the studies of the heritability of IQ, these five have more or less serious methodological problems, and no positive conclusions can be reached using them. The point is not that they prove a genetical identity between races, which they certainly do not, but that there is no evidence for any genetic difference in IQ score. The first four studies, the only ones then available, were reviewed in a report that was meant to be the final judicious word from the American social science establishment, "Race Differences in Intelligence," under the auspices of the Social Sciences Research Council's Committee on Biological Bases of Social Behavior.[43] It is characteristic of the deep ideological commitment of American social science to a hereditarian point of view that the results were characterized as showing that

> Observed average differences in the scores of members of different U.S. racial-ethnic groups on intellectual ability tests probably reflect in part inadequacies and biases in the tests themselves, in part differences in environmental conditions among the groups, and in part genetic differences among the groups. . . . A rather wide range of positions concerning the relative weight to be given to the three factors can reasonably be taken on the basis of the current evidence, and a sensible person's position might well differ for different abilities, for different groups, and different tests.

Precisely how a "sensible person" could reasonably take the position that the observed difference between U.S. racial-ethnic groups is partly genetic, on the basis of the evidence presented, we are not told. Nor is it revealed by this disingenuous summary that, where differences were seen in those observations, they were in favor of *blacks*.

The evidence on cross-class adoptions is sparse. In one sense, adoption in general is cross-class because adopting parents as a group are richer, better educated, and older than the biological parents; and, as we have seen, adopted children have significantly raised IQs.

The study conducted in France by Schiff et al.,[44] however, was designed especially to test the effect of class. The investigators located thirty-two children who had been born to lower-working-class parents, but who had been adopted before six months of age by upper-middle-class (or above) parents. They also located twenty *biological* siblings of the same children. These siblings had been reared by their own working-class mothers. Thus, the two groups of siblings were genetically equivalent but had experienced quite different sorts of environments. The adopted children, by school age, had an average IQ of 111, 16 points higher than that of their stay-at-home siblings. Perhaps more important, 56 percent of the stay-at-homes had failed at least one year in the French school system, compared to only 13 percent of the adopted children.

We should recall that the title of the article by A. R. Jensen that rekindled interest in the heritability and fixity of IQ was "How Much Can We Boost IQ and Scholastic Achievement?" The answer, from cross-racial and cross-class adoption studies, seems unambiguous: As much as social organization will allow. It is not biology that stands in our way.

CHAPTER SIX

THE DETERMINED PATRIARCHY

"Is it a boy or a girl?" is still one of the first questions asked about any newborn infant. This question marks the beginning of one of the most important distinctions our culture makes between people, for whether a child is a boy or a girl is going to make a profound difference to its subsequent life. It will determine its life expectancy. On average, slightly more boys are born than girls; at all ages males have a somewhat greater chance of dying than females; in Britain and the United States at the moment the average male life expectancy is about 70 years, while that of females is about 76. This means that most elderly people are women—more than three women to every man in the 85+ age group, for instance.

In Western society today, on average, men are taller and heavier than women. They have larger brains, compared to women, though not when considered in proportion to body weight. Men and women show differential susceptibility to many diseases, quite apart from the obvious, reproductive ones: men suffer more frequently in our culture from a variety of circulatory and heart diseases and some cancers; women are more likely to be diagnosed as psychiatrically disturbed and to be drugged or institutionalized as a result. Men are physically stronger in terms of performance on the sports field or track. Even though a high proportion of women are in paid labor outside the home, the jobs they do tend to be different from those of men. Men are more likely to be cabinet ministers or parliamentarians, business executives or tycoons, Nobel Prize–winning scientists or fellows of academies, doctors or airline pilots. Women are more likely to be secretaries, laboratory technicians, office cleaners, nurses, airline stewardesses, primary school teachers, or social workers.

And these differences in "chosen" profession are mirrored in school performance and the behavior of children at an early age. Boys play with cars and construction sets and cognitive board games; girls with dolls, shops, nurse's uniforms, and home cooking sets. Girls expect to be primarily homemakers, boys to be breadwinners. Fewer girls at school study technical subjects, science, or metalwork; fewer boys study home economics. After adolescence, girls perform worse than boys at math.

All these are current "facts," objectively ascertainable statements about our present society at this time in history. Some are seemingly facts about biology, some about society, and some about both. But how are they to be understood? What are their implications, if any, for assessing the limits to social plasticity? More than almost any other social "fact" with which this book deals, "facts" about differences between men and women in society—*gender* differences—are seemingly naturalized as manifestations of essentially biological *sex* differences, so apparently obvious as to be beyond question. And indeed for many men, such assumptions—which imply that the current division of labor between the sexes in our society (a *social* division of labor) is merely a reflection of some underlying biological necessity, so that society is a faithful mirror of that biology—are extraordinarily convenient.[1]

That we live in a society characterized by differences of status, wealth, and power between men and women is abundantly clear. Just as contemporary Western society is capitalist in its form, so it is also patriarchal.[2] The division of labor between men and women is such that within productive labor men tend to predominate in the more powerful and better paid, more dominant jobs, women in the less powerful, more poorly paid, and more subordinate ones. One whole category of labor—reproductive, or caring labor—is assigned largely if not exclusively to women. Reproductive labor does not just involve the biological labor of childbearing but also the task of organizing the male worker's feeding, clothing, and domestic comfort, nursing him when sick, and so forth. In addition, there is the crucial educational-ideological role of preparing the next generation for *its* productive-reproductive labor by teaching, training, and the transmission of values. That is, either in the home or in the paid sector of the economy, women are disproportionately employed as the preparers of food, the minders and teachers of children, and the nurses of the sick. This division of labor is a feature not merely of Western capitalist societies but also, in varying degrees, of societies that have gone through revolutionary struggle—from the Soviet Union to China, Vietnam, and Cuba.

Why does the patriarchy persist? One possibility is that it is a historically contingent form of social organization, preserved by those who benefit from it, a consequence of human biology, just as any other social form is a consequence of that biology but only one of a range of possible social organizations available to us. Others would argue, by contrast that it is an inevitable product of our biology, fixed by the biological differences between men and women, determined by our genes.

In response to the upsurge of the feminist movement, its social and political demands, its burgeoning theoretical writing of the past decade, biological determinism has stood firm in claiming that occupying leadership roles in public, political, and cultural life goes with being male as much as having a penis, testicles, and facial hair. Women's intrusions into traditionally male preserves are fervently opposed. When simple votes of exclusion to professional male domains fail, biology is invoked. Women should not be bank managers or politicians, for example. As one American doctor put it:

If you had an investment in a bank, you wouldn't want the president of the bank making a loan under those raging hormonal influences at that particular period. Suppose we had a president in the White House, a menopausal woman president who had to make the decision of the Bay of Pigs, which was of course a bad one, on a Russian contretemps with Cuba at that time?"[3]*

Indeed, there is a danger in women even being in any senior position in business. A front-page headline in the *Wall Street Journal* informs us that "firms are disrupted by [a] wave of pregnancy at the manager level . . . problems are more widespread these days because more women hold high level jobs and because pregnancies are increasing among those over 30."[5] And the article goes on to explain that male executives have to work harder at short notice because of inconsiderate attacks of pregnancy on the part of their female colleagues. Moral: Women should only be in jobs in which they can be easily replaced, such as a production line or typing pool. Ignored, of course, in such an account of the problems caused by pregnant female executives is the inconvenience resulting from the high and *unplanned* rate of coronary heart disease among male executives, which must be at least as disruptive. But that is normal.

And the conclusion is of course clear: For women to work outside the home is a mistake; bad for the economy, which must then provide and pay for welfare services which would otherwise be supplied by women's unpaid labor; and against nature, which decrees that the man should be the breadwinner, the woman the raiser of children. New Right ideology is explicit on this point, despite the fact that in both Britain and the United States at least one in six households is solely dependent on the earnings of a female breadwinner.[6]

Resurgent New Right thinking rationalizes this opposition to feminist demands still further. For Britain's National Front, the natural place of women is as much tied to *Kinder-Küche-Kirche* as was that of their Nazi forebears. This view was echoed by Enoch Powell, an M.P. in Britain's Parliament, in debate on the Thatcher government's Nationality Bill (which creates categories of British citizenship roughly designed to make a significant proportion of black British second-class

---

*This hormonal naturalization has its reverse in the recent (1981) acquittal of two women on murder charges in Britain on the ground that they killed while suffering from "premenstrual tension"—a decision welcomed by some feminist voices as liberatory and condemned by others as firmly biologistic, freeing these women while by implication oppressing all.[4]

citizens). Mr. Powell, proposing that British citizenship should be passed on only through the father, explained that plans to let a child claim nationality through its mother was "a concession to a temporary fashion based upon a shallow analysis of human nature. . . . Men and women," he went on, "have distinct social functions, with men as fighters and women responsible for creating and preserving life; societies can be destroyed by teaching themselves myths which are inconsistent with the nature of man [sic]."[7]

For biological determinists, then, gender divisions in society do indeed map onto biological, sexual ones. Not only is the division of labor given by biology, but we go against it at our peril, for it is functional. Society needs both dominant, productive men and dependent, nurturative, and reproductive women.

The biological determinist argument follows a by now familiar structure: It begins with the citation of "evidence," the "facts" of differences between men and women such as those described in the first paragraphs of this chapter. These "facts," which are taken as unquestioned, are seen as depending on prior psychological tendencies which in turn are accounted for by underlying biological differences between males and females at the level of brain structure or hormones. Biological determinism then shows that male-female differences in behavior among humans are paralleled by those found in nonhuman societies—among primates or rodents or birds, or even dung beetles—giving them an apparent universality that cannot be gainsaid by merely wishing things were different or fairer. Biological laws brook no appeal. And finally, the determinist argument endeavors to weld all currently observed differences together on the basis of the now familiar and Panglossian sociobiological arguments: that sexual divisions have emerged adaptively by natural selection, as a result of the different biological roles in reproduction of the two sexes, and have evolved to the maximal advantage of both; the inequalities are not merely inevitable but functional, too.

In the present chapter we will review these apparently scientific claims to explain the current gender divisions in society and will show that they represent a systematic selection, misrepresentation, or improper extrapolation of the evidence, larded with prejudice and basted in poor theory, and that, far from accounting for present divisions, they serve as ideologies that help perpetuate them. As for biological explanations of the differences in IQ performance between races and social classes, the objective of biological explanations of present sex roles is to justify and maintain the status quo.

The Status of the "Facts"

The persistent claim of biological determinist thinking is that the social structure of contemporary Western society mirrors general social structures that are universal. At worst, because of "unnatural" liberal and radical pressure, we have fallen from some prior state of social Darwinist grace. At best, we are what we must be. Hence "facts" of the type of the first paragraphs of this chapter are given a spurious universality. Take job distribution. The present universality of women in office jobs masks the fact that until the early part of this century clerking was an exclusively male preserve and efforts were made to keep women out of office work.[8] "Biological" reasons were advanced then as to why they were unfit for such labor, just as in 1978 the journal *Psychology Today* could report that "as women in general are superior in fine coordination and the ability to make rapid choices, they may for example be faster typists than men."[9] Temporal shortsightedness is matched by geographical shortsightedness; for instance, although it might seem natural that men dominate medical practice in the United States, this situation is reversed for family doctors in the Soviet Union, where the majority are women. (Of course, there family doctoring has a lower status and lower pay than in the United States, but that is a different point.)

The particular dating patterns, sexual practices, and fashion styles of 1950s American teenagers are among those most strikingly universalized by biological determinism. In a well-known study of girls who had been "masculinized" by exposure *in utero* to androgenic steroids administered to their mothers, Money and Ehrhardt define the femininity of their subjects by specific criteria, including whether they show a liking for jewelry, wear pants, manifest so-called tomboyish behavior, or are more concerned about a future career than about a romantic marriage.[10] This point doesn't merely embrace the ideology of the women's magazines that provides a set of acceptable standards —stereotypes; it ignores the existence of societies in which women wear pants, or in which men wear skirts, or in which men enjoy and appropriate jewelry to themselves. The girls are being judged by Money and Ehrhardt by how well they conform to the stereotyped local image of femininity. They are shown to have been mildly willing to reject these forms—though they still expected to marry and be mothers. And this rejection—among girls who were aware of the ambiguities of their own gender labeling and the unusual attention that the researchers were paying to them as opposed to their peers—is supposed to be expressive of some universal biological determination.

Naiveté in the description of human social and sexual arrangements displayed by biological determinists also characterizes the attention that sociobiologists such as Wilson, Van den Berghe, and others have paid to a phenomenon they regard as universally human—the "incest taboo." Yet inspection of the sociological literature, if nothing else, would soon have told them that even in present-day Western societies laws against incest are no barrier to its substantial incidence.[11]*

This kind of thinking is redolent as much of a social as of a sexual

---

*The incest taboo is one of the odder sociobiological stories.[12] The argument begins from the genetically correct statement that brother-sister matings are likely to increase the number of offspring with disabling or deleterious double recessive genes, and are therefore eugenically unfavorable. It would therefore be an adaptive advantage for such close-kin matings to be avoided. Sociobiology claims that this is indeed the case among both humans and nonhumans. The mechanism whereby we and other organisms recognize one another's genetic relatedness and hence sexual availability is unspecified; one suggestion is that the rule is "Do not mate with someone you have been brought up with." The nonhuman evidence is at best fragmentary; the prediction seems to be supported by observations of some baboon populations, and by unfortunate extrapolations from the behavior of new-hatched Japanese quail; but the common observation of fairly indiscriminate mating among domestic or farmyard animals is met by the bland assurance that such species have been peculiarized by human intervention. So far as humans are concerned, social rules of permitted and forbidden mating patterns in a large number of different societies tend to be cited. Yet even if it were true that there was a universal incest taboo that forbade genetically close marriages (which there is not), it is not possible to map social definitions of kin directly onto genetic ones; and even if it were also true that this taboo was followed in practice (which it is not), the argument makes no sociobiological sense. For if the "taboo" is indeed genetically prescribed, what need is there for mere social legislation to enforce it? A natural repugnance should require no legal shoring up in this way. Unless of course it is not that our genes inhibit us from copulating with our siblings but instead induce us to pass laws regulating such copulation.

chauvinism, a chauvinism that knows nothing except the stereotype of its own society within a very sharply delimited class border. It is a narrowness that knows neither sociology, history, nor geography. Social universals then appear to lie more in the eyes of the biological determinist observer than in the social reality that is being observed. But this is interestingly true of the apparent biological universals as well. Some are straightforward. The fact that today the life expectancy of females exceeds that of males in advanced industrial societies is very sharply affected by the dramatic decline of the death rate in or around childbirth that was so characteristic of women everywhere until the present century. Morbidity statistics show comparably rapid changes. In the United States and England women are steadily catching up with men in the rate of deaths from lung cancer and coronary thrombosis, for instance. Less obvious are such phenomena as the secular decline in sexual dimorphism in height that has been recorded over the past century. The average male-female height difference was substantially greater a century ago than it is today in advanced industrial societies. Or take the relative performance of men and women in sport. What would have been perceived only a few decades ago as a natural and inevitable difference between men and women has steadily been eroded. Dyer looked at average male-female differences in track athletics, swimming, and time trial cycling between 1948 and 1976 and showed that in each of these three sports women's performances in relationship to men have continuously improved, and that if these changes are continued, the average female performance will equal that of males for all events currently competed for by both sexes sometime during the next century.[13]

But how important are averages anyway? The fact that today on average men are taller than women does not deny that many women are taller than many men. Average statements about populations are only made *post hoc*, that is, after we have decided on the definition of the populations being described. Thus, before we can describe differences between men and women we have to define the two populations —male and female—to be compared. It is just this dichotomization that is under discussion, though, and which we are urging cannot merely be dismissed as "natural."[14] If the dichotomization masks such overlap yet serves the social functions of pushing people into one or other of two boxes labeled "man" or "woman," then attempts to pontificate about the nature and origins of the differences are in deep trouble. "Average" statements are powerful, but they are not necessarily the most helpful ways of describing phenomena. Furthermore, they run the danger of becoming self-fulfilling. If there are average stereotypes to which girls and boys are encouraged to conform—so that boys practice being "masculine," girls practice being "feminine"—the stereotypes perpetuate dichotomies and further enhance the appearance of being "natural."

The next step in the determinist proof of these social "facts" is to map observed social divisions onto individual psychological ones. According to their arguments, when one examines the psychology of either sex we find that women excel at certain tasks, men at others. Note that differences between the sexes in average IQ cannot be claimed, because the standard IQ tests as developed in the 1930s were carefully balanced to eliminate any sex differences that the earlier version of the tests had shown. Thus an earlier generation of determinists has neatly removed this particular weapon from patriarchy's ideological armamentarium. Fairweather summarized the received wisdom of the psychology of sex differences as follows:

Females have been seen . . . as more receptive . . . within the tactile and auditory domain, although retaining particular high class discriminatory abilities such as that involving face recognition. . . . Emotionally more dependent, they are "sympathetic" both in nature and nervous system. As a result less exploratory, they fail to develop the independence of immediate surrounds necessary for orientation in large spaces, or the manipulation of more immediate spatial relations. Cerebrally, they live with language in the left hemisphere. Males on the contrary are characterised as brashly visual, preferring simple responsive stimuli, responding best with grosser movements; fearless and independent; parasympathetic and right hemispheric; and ultimately, successful.[15]

Thus men and women have different success rates in different jobs because they are doing what comes naturally.*

*Such naturalizing is not confined to "obvious" reactionaries. William Morris, in his anarchical vision *News From Nowhere*, describes his free society as one in which the women cook and wait upon the men at table because this is what they "naturally"

According to Maccoby and Jacklin, girls have greater verbal ability than boys, and boys excel in visual-spatial (mechanical aptitude) and mathematical skills and are more aggressive than girls.[16] The consequence, according to psychologist Sandra Witelson, is that there may be fewer female architects, engineers, and artists

because such professions require the kind of thinking that may depend on spatial skills . . . in contrast women performers (vocalists, instrumentalists) and writers are less rare. This may be because the skills involved in these talents may depend on functions women do well—linguistic and fine motor coordination.[17]

Job choices in a free society are thus merely the reading out of individual preferences—ontologically prior personal decisions based on innate psychology. The social forces driving particular "choices"—the directing influences of school and families, or the male exclusion of females from particular trades and professions—are all irrelevant. That in the United States and the United Kingdom adolescent girls do worse than boys at math is quickly taken as evidence "that sex differences in achievement in and attitudes toward mathematics results from superior male mathematical ability, which may in turn be related to greater male ability in spatial tasks."[18]

Ignoring the social and cultural pressures driving the sexes in different directions, the consistently reported driving out or putting down of girls who show interest in math leads directly to the biological explanation.[19] To come back to Witelson's examples, Virginia Woolf pointed out a long time ago that in a society in which women are denied even the privilege of space—a room of one's own—almost the only permissible skills become those that do not demand privacy or space: A writer's notebook is transportable, a painter's canvas or architect's drawing board not so. And while women's "accomplishments" are praiseworthy, real expertise that might challenge the male or take time away from the crucial reproductive role is not. (The new feminist

enjoy. However, in Utopia men recognize the skills involved in these activities and respect the women for them. Male black-power spokesmen have been known to take a similar position. At the 1981 Labour Party Conference, when the chairman thanked the women for their tea making, he was successfully challenged by feminists under the slogan "Women make policy, not tea."

scholarship has chronicled an entire history of nineteenth-century medical men and psychologists insisting on the antithesis between creative work—for instance in scholarship or science—and reproduction. Women who studied would damage their essential reproductive capacity.)[20]

But how valid are the psychological claims made by Witelson and others? Are these "differences" real, and if they are, can one ascribe causes to them? These days most researchers recognize that the differences observed between men and women, or even schoolchildren, represent the outcome of an inextricable interplay of biological, cultural, and social forces with genotype during development. So the tendency has been to seek methods to research psychological traits in younger and younger children, or even in newborns. Reviews and popular books[21] claim that even here differences are found— differences in crying, in sleep patterns, in smiling, in latencies to particular responses—which lay the basis for what is to come. Yet in an exhaustive review of the literature on sex differences and performance among newborns, Fairweather was able to conclude that despite persistent claims to the contrary:

In childhood we are left with, at most, a female propensity for precise digital movement; and on the same continuum, a male propensity for activity demanding the usage of larger musculatures and certain spatial (body-orientational) abilities which these may subserve. The rest is dilemma.[22]

In slightly older children there are

no substantial sex differences in verbal sub-tests of IQ tests; in reading; in para-reading skills (Cross-Modal matching); and in early linguistic output; in articulatory competence; in vocabulary; and in laboratory studies of the handling of verbal concepts and processing of verbal materials.[23]

Differences emerge only later when the "sudden polarization of abilities in adolescence" occurs.

So the actual evidence for sex differences in cognitive behavior among infants is slight. But even were there such evidence, what would it prove? Is it that by going back to infancy one can study "pure, biologically determined behavior" uncontaminated by culture? The answer is no. A child can only develop in an environment that includes the social from the earliest postnatal moment.* Babies interact with their caregivers; they are held, clothed, fed, cuddled, and talked to. Parents are said to hold and speak to their male babies differently from their female babies, quite apart from dressing them in blue or pink.[25] All cultures must generate expectations of behavior among parents and hence ensure that certain types of behavior are going to be consciously or unconsciously reinforced or discouraged from the beginning. This would be the case whether the infant was cared for by a biological parent or by surrogates. We are not trying to transfer "blame" to the mothers. The point is that the determinants of behavior are irredeemably interactive and ontogenetic. However young the child that is being studied, its behavior must be the product of such interaction. To argue that one can divide behavior chronologically into a portion given by biology and another given by culture is to fall into a reductionist trap from the start. This is not to discount the importance of studying the development of behavior in young children, which is among the most fascinating areas of human ethology. We insist, however, that such studies do not ask naively reductionist questions of their subject; what is needed is as rich and interactive a methodology as the varied development of human infants themselves.

But apparent psychological differences between the sexes are just the starting point for the biological determinist argument. If there are such differences, they must, the argument runs, reflect underlying differences in brain biology. Somehow, if the differences can be grounded in biology they are seen as more secure from environmental challenge. Again we should emphasize that as materialists we too would expect to find that differences in behavior between individual humans will prove to be related to differences in the biology of those individuals. Where we differ from biological reductionism is in refusing to accept an argument that says that the biological difference is primary and causative of the "higher level" psychological one; *both* are different aspects of the same unitary phenomenon. Differences in the social environment of an individual during development can result in changes in the biology of brain and body just as much as in behavior.

*We do not discuss here the effects of the prenatal environment on development, important though these may be.[24]

So to show that there are differences between the brains of males and females on average says little about either the causes or the consequences of such differences.

But are there differences? Certainly the belief in them goes back a long way. Nineteenth-century anthropologists were obsessed with the question of the relationship between intelligence and brain size. Just as they were convinced that the white brain was better developed than the black, so the male was superior to the female. Male brains were heavier, as the neuroanatomist Paul Broca pointed out, but there were also differences in structure. According to the anthropologist McGrigor Allan in 1869, "The type of the female skull approaches in many respects that of the infant, and still more that of the lower races."[26]*

Much was made of the "missing five ounces" of the female brain until it was realized that when brain weight is expressed in relation to body weight the difference disappeared or even reversed. This led to further devices such as comparing brain weight to thigh-bone weight or body height.[27] Attention was diverted to brain regions—frontal or temporal lobes, for example—as the seats of differences. It was left to Alice Leigh, a student of Karl Pearson, in 1901, using new statistical methods, to conclude that there was no correlation between skull capacity and hence brain weight and "intellectual power."[28]

For many years subsequently neuroanatomy and neurophysiology measured no differences between male and female brains. Only with the emergence of new methodology in anatomy, physiology, and biochemistry in the 1960s and 1970s (and the rise of the new biological determinism) has the question become actively asked once more. Most

*The juxtaposition of sexism and racism was a characteristic feature of nineteenth-century biological determinist thinking. Charles Darwin commented that "some at least of those mental traits in which women may excel are traits characteristic of the lower races."[29]

For the French craniologist F. Pruner, "The Negro resembles the female in his love for children, his family and his cabin . . . the black man is to the white man what woman is to man in general, a loving being and a being of pleasure.[30]

The theme runs through much of nineteenth-century evolutionary and anthropological writing and finds a curious time echo in the contemporary suggestion by Arthur Jensen that because (he claims) spatial perception is a sex-linked ability, it can be used effectively to study the relationship of black-white gene mixing on racial differences in intelligence.[31] *Plus ça change.*

attention has been devoted to the claims that there is a difference in so-called lateralization between male and female brains. As a structure the brain divides neatly into two virtually symmetrical halves, like the two halves of a walnut, the left half (hemisphere) being broadly associated with right-side-of-body activity, the right half with left-side activity. This symmetry is incomplete, however. It has been known since Broca's day, in the nineteenth century, that speech and linguistic functions are, for most people, located in a region of the left hemisphere—parts of the temporal lobe. Hence left-hemisphere strokes or thromboses affect speech, whereas equivalent right-hemisphere brain damage generally does not. The regions of the temporal lobe in the left hemisphere that apparently accommodate speech are anatomically larger than the corresponding right-hemisphere regions.

Evidence for sexual dimorphism in hemispheric size in humans has begun to appear and seems more securely grounded than earlier claims for meaningful differences in overall brain size. How such differences arise is unclear: one possibility offered by Geschwind and his colleagues is that there are interactions of the fetal brain during development with hormones such as testosterone, of which more later. It is argued that testosterone slows the growth of the left hemisphere relative to the right.[32] As is characteristic in such analyses, animal data is cited to support the case in humans; thus part of the right cerebral cortex of the rat is thicker in males, while the corresponding part of the left hemisphere is thicker in females, and these differences are modified by experimentally changing the hormonal balance of the animals in infancy.

There are two major problems with interpreting the significance of such observations. The first is that of extrapolating from nonhuman to human brains. While the nerve cells—the basic units that make up the brain—and the way these cells individually work are virtually identical in organisms as diverse as sea slugs and humans, the number of cells, their arrangement, and their interconnections differ dramatically. Insects and mollusks have a few tens or hundreds of thousands of nerve cells in their central ganglia, a rat or a cat may have hundreds of millions in its brain, and humans have between ten and a hundred billion in theirs, each communicating with its neighbors by up to a hundred thousand connections. Brain weight for body weight, only a few primate species and dolphins approach this order of complexity. Further, in organisms with less complex brains most of the neural pathways are laid down—genetically specified—to form rather rigid and preprogrammed connections. This invariance gives such organisms a comparatively fixed and limited behavioral repertoire.

By contrast, the human infant is born with relatively few of its neural pathways already committed. During its long infancy, connections between nerve cells are formed on the basis not merely of specific epigenetic programming but in the light of experience. The microchips in a pocket calculator and a big general-purpose computer may be similar in composition and structure, but one is a limited, dedicated piece of machinery with a fixed repertoire of outputs, while the other is vastly variable. Homologies of structure between animal and human brains are of interest, but one cannot ascribe homologies of meaning, still less identities, to their outputs on this basis alone. For instance, there is a marked sexual dimorphism in the brains of certain species, notably songbirds. The male canary has a concentration of nerve cells in a particular brain region that is lacking in the female and is associated with the generation of its song, the development of which is hormone dependent.[33] This brain region is relatively smaller in the female canary. This does not allow us to predict ways in which postmortem analysis of brains would have found differences between canaries and Maria Callas, however. Nor does it allow us to deduce where in Callas's brain her singing capacity was located. Homologies of structure between species do not mean homologies of function.

Biological determinism makes great play of the evolutionary origins of the human brain, in which certain deep structures can be shown to have first evolved in our reptilian ancestors. Maclean has spoken of the "triune brain,"[34] whose three broad divisions can be derived from humanity's reptilian, mammalian, and primate forebears. But it is absurd to conclude, as some determinist arguments seem committed to doing, that with part of our brain we therefore think like snakes.[35] Evolutionary processes are parsimonious with structures, pressing them constantly to new purposes rather than radically abandoning them. Feet become hooves or hands, but we do not therefore conclude that hands behave in hooflike ways. The human cerebral cortex evolved from a structure that in more primitive-brained forebears was largely the organ of olfaction. This does not mean that we think by smelling. (The question of homology is further discussed in Chapters 9 and 10.)

Localizing emotions and behavioral capacities has been the sport of determinism since the days of phrenology. Yet while it is clearly true that we can say of particular brain regions that they are *necessary* for given behaviors (or their expression), there is no region of the human brain of which we can say that it is *sufficient* for such functions.* One cannot see without eyes; one still cannot see with eyes but without the vast regions of both halves of the brain to which the eyes are connected. And the property of perception—the analysis of visual information—is not localized either to the eyes or to any particular constellation of cells within the brain; rather, it is a property of the entire eye-brain system, with its interconnecting web of nerve cells.

So the fact of anatomical differences in brain structure between males and females, itself no more and no less interesting than the fact of anatomical differences in genitalia between the two sexes, does not permit us to draw conclusions about the biological substrate or innateness of behavioral differences. Just what hemispheric differences mean is simply unknown, despite the literature of hemispheric specialization that has grown up in the past decades. It has been suggested, for instance that paralleling the linguistic skills of the left hemisphere are spatial ones in the right; that the left hemisphere is cognitive, the right affective; that the left is linear, digital, and active, while the right is nonlinear, analogic, and contemplative; that the left is Western, the right Eastern. One prominent Catholic neurophysiologist has placed the seat of the soul in the left hemisphere. Hemispheric specialization has become a sort of trash can for all sorts of mystical speculation.[36]

And to this list of speculative differences have now been added sex differences. If men have greater spatial perceptual abilities, and women better linguistic skills, one might anticipate that men would be more "right-hemispheric," women more "left-hemispheric." But this won't do. Men are also cognitive (said to be a left-hemisphere function) and women affective (said to be a right-hemisphere function). To preserve male cognitive and spatial preeminence and yet map these onto brain structures the male brain must be described as more lateralized—each half does its own thing better; while women are less lateralized—the two halves of the female brain interact more than do those of the male. Hence men can do different types of things simultaneously whereas women can only do one thing at a time without getting confused (it is not true that Gerald Ford was female, however).

*This theme, of the localization fallacy, occurs again in relation to the "site" of violent behavior, discussed in Chapter 7.

The possibilities of stereotypic speculation based on differences in lateralization are obviously immense. Witelson expresses the confusion neatly:

For example men are superior on tests of spatial skills and tend to show greater lateralisation of spatial function to the right hemisphere. Here greater lateralisation seems to be correlated with greater ability. However, in the case of language, women, in general, are superior to men, who show greater lateralisation of language skills to the left hemisphere. Thus with language greater lateralisation may be correlated with less ability.[37]

Witelson's enthusiasm for overinterpretation of data is not unique. Even some feminist writers have adopted the lateralization argument and made it over for their own purposes. In conformity with one strand of feminist writing which, like that of masculine biological determinists, argues for essential differences between the ways in which men and women think and feel but rejoices in the superiority of the female mode, Gina argues that women should welcome the intuitive and emotional strengths given by their right hemisphere, in opposition to the overcognitive, left-hemisphere dominated, masculine nature.[38] While we would agree that the peculiarly reductive or objectivist nature of scientific knowledge as it has developed in the context of a patriarchal and capitalist society is to be opposed, we would not accept that reductionist science is innately wired into the masculine brain.

The truth of the matter is that while the evidence for hemispheric differentiation and specialization of function is among the most intriguing of the developments in human neurosciences of the past decade, its relationship to individual differences in behavior is quite unclear, except in the case of brain damage or disease in adults, where the capacity for the plastic recovery of function is very limited. (Children show much more plasticity.) Differences in lateralization, if they exist, are not explanations for social divisions, though they provide a fertile ground for biological determinist imagination.

If biologically determined male-female cognitive differences divorced from the social framework begin to dissolve on inspection, there is one difference on which all biological determinists are agreed: Men and boys are more aggressive than women and girls, a difference that appears at an early age, when it manifests itself in an activity called rough-and-tumble play* and continues into adulthood, where it is expressed as a need or tendency to dominate. Men may not be better than women at any particular task, but they are prepared to push and shove their way to the top more aggressively. The argument received its fullest expression in the mid-1970s in a book by Steven Goldberg, *The Inevitability of Patriarchy.* [39]

Goldberg's argument is engagingly direct: Wherever one looks, in all human societies throughout all history, there is a patriarchy. "Authority and leadership are and always have been associated with the male in every society" (page 25). Such universality must imply "the strong possibility that these may be inevitable social manifestations of human physiology" (page 24). Attempts to create a different society must fail, as "the inexorable pull of sexual and familial biological forces eventually overcome the initial thrust of nationalistic, religious, ideological or psychological forces that had made possible the temporary implementation of Utopian ideas" (page 36). Men always have the high-status roles, not because women can't do them but because they are not "for psychophysiological reasons . . . as strongly motivated to attain them" (page 46).

The magic lies in a "neuroendocrinological differentiation" (page 64), which gives the male a greater tendency to dominate. Men will dominate, whatever behavior this might require: "Fighting, kissing babies for votes, or whatever . . . it is not possible to predict what the necessary behavior will be in any given society because this will be determined by social factors, but whatever it may be it will be manifested by males" (page 68). Domination is ensured in groups and in dyads (i.e., men want to boss other men and their female sexual partners and children). The neuroendocrinology must be very flexible if it can generate such varied expressions, of course. It is a very bold neuroendocrinologist who would want to argue that the hormonal features involved in kissing babies are identical with those involved in fighting, but Goldberg is not deterred. Everything lies in the hormones, which at a particular phase of development "masculinize" the fetal brain. The magic hormone itself is testosterone, produced by the testes, seen as the

*Rough-and-tumble play is supposedly more frequent not merely in young male humans than in young females but also in males of several other mammalian species. However, its relationship to aggression is largely inferential.

"male" hormone, whose presence around birth probably produces some change in brain mechanisms with lasting subsequent effects.*

And if men have this Nietzschean will to dominate, what do women have in its stead? Goldberg waxes poetic. Women's hormones provide them with "a greater nurturative tendency (i.e., they react to a child in distress more strongly and more quickly than do males)" (page 105). Women's role is that of "the directors of society's emotional resources . . . there are few women who can outfight [men] and few who can out-argue but . . . when a woman uses feminine means she can command a loyalty that no amount of dominant behavior ever could." What a touching picture of Goldberg's vulnerability to seduction is thus revealed! So like the home life of our own dear nuclear family. Go against it at your peril. Women should not "deny their own nature . . . argue against their own juices" (page 195). In every society a basic male motivation is the feeling that the women and children must be protected. "But the feminist cannot have it both ways: if she wants to sacrifice all this, all that she will get in return is the right to meet men on male terms. She will lose" (page 196). For Goldberg, then, the interplay of "male" and "female" hormones with the brain, starting early in development, is the key to the sexual universe. However, when one comes to sort out the biology from the rhetoric, the magic power of these baby-kissing and fighting or nuturant and juicy hormones seems to fade.

The Biology of Sex

What lies behind the Goldbergian thesis of "male" and "female" hormones? A digression on the biology of sex (as opposed to gender) differences in humans is necessary here. Human sexual differentiation in embryonic development begins with the influence of the chromosome carried by the sperm. Of the twenty-three pairs of chromosome in each body cell of a normal person, twenty-two are autosomes—nonsexual chromosomes—present in two copies in either sex. The twenty-third pair are the sex chromosomes. Normal females carry a

*Actually, Goldberg's evidence for the effect of testosterone on brain mechanism is derived in large part from studies on mice and rats.

pair of X chromosomes, while males have one X and one Y chromosome. This is achieved because all ova have a single X chromosome, and sperm may carry *either* an X *or* a Y; hence the fertilized egg that results from mating will be either XX or XY depending on which sperm fertilizes the eggs. At first sight, sex differences depend on the differences between an XX and an XY. For certain single characters this may largely be the case. For instance, the absence of the second X in males means that some deleterious recessive genes whose effects would otherwise be masked are expressed; females carry traits such as color blindness or hemophilia, but they are expressed in males, as sex-linked traits. But of course during development genes interact with one another—or rather the protein products of one gene interact with the protein products of another—in complex ways, and hence autosomal and sex chromosome products will be involved mutually in the development of the organism.

Attempts are sometimes made to infer the consequence of possession of X or Y chromosomes from the study of individuals with rare chromosomal abnormalities. For instance, in Turner's syndrome one of the sex chromosomes is absent (XO); in Klinefelter's syndrome there is an extra X (XXY). Males who carry an extra Y (XYY) have sometimes been described as "supermales," and efforts have been made to prove that they have higher levels of "male" hormones or are unusually aggressive or criminally inclined. Despite a flurry of enthusiasm for such claims in the late sixties and early seventies, they are now generally discounted.[40]

In any event, such inferences about the role of the Y chromosome in normal development are always doomed to failure. The presence of an additional chromosome produces effects that are not merely additive or subtractive to a normal developmental program; rather, such a presence throws the whole program out of kilter. Down's syndrome,* for instance, is a chromosomal disorder in which there is an extra autosomal chromosome (trisomy 21), but the result of the addition is to produce an individual with a wide range of defects—retarded men-

---

*Down's syndrome used to be called mongolism, a reference to the naive racism of nineteenth-century clinicians who viewed idiocy in the "white races" as reflective of "throwbacks" to "more primitive" black, brown, and yellow races. Of the various terms used to classify "idiocy" in this racial typology, only "mongolism" survived for any length of time.

tal, motor, and sexual development, low IQ test scores, and some disordered physical characteristics often including webbed fingers and toes. But the disorder has some positive features as well. For example, Down's syndrome children are often conspicuously happy and friendly, with "sunny" dispositions. We should not be surprised at such complex phenotypic consequences.

The Y chromosome does play an important role during normal development in the expression of male physiological and morphological characteristics, particularly for the differentiation of the testes. During embryonic development, the primitive sex gland that develops during the first few weeks requires the presence of a Y chromosome in order to differentiate into the testis. In both sexes, hormonal secretion begins to occur. Now, contrary to the impression conveyed by Goldbergian hormonal determinism, and indeed to the naming of hormones as estrogens and androgens, such sex hormones are not uniquely male or female. Both sexes secrete both types of hormone; what differs is the ratio of estrogen to androgen in the two sexes. Hormones (gonadotropins) from the pituitary—a small gland at the base of the brain—regulate hormone release from both ovary and testis, which are then carried to other regions. The presence of both androgens and estrogens (as well as other hormones) seems to be required in both sexes to achieve sexual maturity, and both types of hormone are produced not merely by ovary and testis but also by the adrenal cortex in both sexes. Furthermore, the two kinds of hormone are chemically closely related and can be interconverted by enzymes present in the body. Estrogens were at one time prepared from pregnant mare urine, which excretes the large amount of more than 100 mg daily —a record, as Astwood puts it, "exceeded only by the stallion who despite clear manifestations of virility, liberates into his environment more oestrogen than any living creature."[41] Nor is progesterone—a hormone that affects the development of uterus, vagina, and breasts, which is intimately involved in the processes of pregnancy, and whose rhythmic fluctuations characterize the menstrual cycle—confined to females; it is present in males at levels not unlike those of the preovulatory phase of the menstrual cycle in females. It can be a chemical precursor of testosterone.

So, insofar as sex differences are determined by hormones, they are not a consequence of the activities of uniquely male or female hormones, but rather probably of fluctuating differences in the ratios of these hormones and their interactions with target organs. Genetic sex, determined by the chromosomes, is, during development, overlaid by hormonal sex, shaped by the ratios of androgens to estrogens, and normally, though not always, appropriate to the genetic sex of the individual. Of course the hormones too are produced by gene-initiated processes but are subject to much more environmental change or deliberate manipulation, either by hormone injection or removal of the hormone-producing glands, for instance by castration in animals. Finally, in humans the cultural and social environment of sexual expectations is overlaid yet again on the chromosomal and hormonal phenomena.

From Sex to Gender

There is a conspicuous lack of relationship, in humans, between, on the one hand, levels and ratio of circulating hormones, and on the other, sexual enthusiasms or preferences. In some laboratory animals, notably the rat, there is a relatively straightforward relationship between, say, estrogen and progesterone levels and sexual enthusiasm in the female, so that injection of estrogen induces the female rat to take up a position in which she raises her rump in sexual invitation. But even in the arid environment of a laboratory cage the response of the female to hormone injection depends on her prior experience, and the relationship between hormone levels and sexual activity is even less straightforward in more complicated "real life" environments. In humans the matter is certainly much more complex. Hormone levels are not simply or directly related to either sexual enthusiasm or attractiveness to the opposite sex.

Nor do the hormonal levels or ratios have much to do with the direction of sexual attraction. It has been a popular hypothesis over some forty years that people with homosexual enthusiasms would show levels of circulating hormone more appropriate to the "wrong" sex. Lesbians, it was argued, should have higher androgen and/or lower estrogen levels than heterosexuals.[42] Yet no such relationships exist. Nor would we have expected them to: the very assumption implies a reification and biological reductionism which insists that all sexual activities and proclivities can be dichotomized into hetero- or homo-directed, and that showing one or the other proclivity is an all-or-none state of the individual, rather than a statement about a person in a particular social context at a particular time in his or her history. Of sociobiology's view of the "adaptiveness" of homosexual behavior, more in Chapter 9.

The failure of simplistic attempts to associate hormone levels with sexual enthusiasm or direction led determinists to the assumption that what matters is not so much the adult hormonal level but the interplay of hormones with, for instance, the brain during development—perhaps even prenatally. The role that the steroid hormones play during early development is clearly an important one, not merely in terms of the maturation of the sex organs but also because both estrogens and androgens interact directly with the brain during crucial phases of its development. There are now known to be many regions of the brain —not just those areas of the hypothalamus most directly concerned with the regulation of hormonal release—that contain binding sites at which both estrogens and androgens become concentrated. These sites are present, and hormones become bound to them, not merely prepubertally but even prenatally; and both androgens and estrogens are bound by both males and females, though there are differences in the pattern and scale of binding between the sexes, and differences in the structural effects the hormones have on the cells to which they bind.

Until a few years ago the human brain used to be regarded as "female" until the fifth or sixth week of fetal life, irrespective of the genetic sex of the individual; in normally developing males it was believed to be then "masculinized" as a result of a surge of androgens. But "femaleness" is not simply the absence of "masculinization"; it is now clear that there is also a specific alternative "feminization" process taking place at the same time, though one should be cautious about accepting at face value the unitary nature of processes implied by such terms as "feminization" and "masculinization."[43]

The question is of course not simply whether or not there are hormonal differences between males and females—for clearly there are —nor whether there are small differences, on average, in structure and hormonal interactions between the brains of males and females; clearly this is also the case, though the overlaps are great. The point is the *meaning* of these differences. For the determinist these differences are responsible not merely for differences in behavior between individual men and women but also for the maintenance of a patriarchal social system in which status, wealth, and power are unequally distributed between the sexes. For Goldberg, as the propagandist of patriarchy, there is an unbroken line between androgen binding sites in the brain, rough-and-tumble play in male infants, and the male domination of state, industry, and the nuclear family. Wilson, the sociobiologist, is more cautious: Our biology directs us toward a patriarchy; we can go against it if we wish, but only at the cost of some loss of efficiency.

Differences in power between men and women are thus, for determinism, primarily a matter of hormones. Appropriate doses at a critical phase in development make males more assertive and aggressive; by contrast, they make females less aggressive, or even, in one extraordinary version of the argument, more likely to offer themselves as the victims of male violence. In a book written after a decade of working with battered women in refuges from their violent husbands and lovers, Erin Pizzey claimed that certain categories of both men and women became violence-addicted as a consequence of being exposed to violence as very young children or while still unborn.[44] Their infant brains, she speculates, came to require a regular dose of hormones—which she variously suggests may include adrenaline, cortisone, and the enkephalins which can only be obtained by violent and pain-giving activities. Why on this model it is men who characteristically inflict the pain and women who characteristically are the recipients of it is not made clear. The point is again a structure of argument that (without convincing evidence) traces complex human social interactions to simple biological causes and locates them in a domain so removed from present intervention as to appear inevitable and irredeemable. The fault for male violence lies, in this view, not in the present structure of a society that traps women into relationships of economic as well as emotional dependence, nor in the despair engendered by unemployment and the devastated inner-city environment, but in a biological victimization dependent on the contingencies of hormonal interactions with the brain at or around birth. If the fault is not in our genes, it is at best in our parents; either way the circle of deprivation visits our sins upon our children.

We do not offer to explain away violence against women by replacing biologistic fantasies by crude economic and cultural reductionism. The problem is certainly too serious for that. But the complexity of male domination defies simplistic localization to the effects of hormones in the brain of the newborn. If such a Goldbergian hypothesis were correct we might expect economic and cultural success to be a consequence of individual male aggressiveness. Yet it is not apparent that such individual aggression is the key to climbing the organizational ladder that leads particular men to become successful as entrepreneurs, politicians, or scientists. The range of economic and cultural determinants for such individual successes is far more complex, and we would doubt that we could explain the emergence of a president of the United States or a British prime minister by measuring the circulating androgen levels in the bloodstream of the contenders for such a post —or even by retrospectively speculating on the levels of these hormones in the days or months following their birth. The level of explanation that must be sought lies properly in the psychological, social, and economic domain; biologists are unable to predict the future Ronald Reagans or Margaret Thatchers from any measurement, however sophisticated, of the biochemistry of today's population of newborn infants.

The counterpart to the myth that male domination and the social structure of the patriarchy are given by male hormones is that it is female hormones that produce the nurturative, mothering activity of women—the mothering "instinct." While it is clear that only women can bear children and lactate, and that this very fact is likely to result in a different relationship between a woman and the child she has borne from that which the male parent has with the child, the implications of this, either for adult caregiving to the child or receipt of the care by the child, are quite undetermined. Not merely the range of different caregiving arrangements developed in different cultures but the quick transitions in advice given by experts to women on whether they should leave their children and enter work—as during the Second World War—or return to their "natural" nurturative activities bear witness to the fact that child-care arrangements owe more to culture than to nature. To recognize the centrality of reproductive, caring labor to human society, the role of mothering,[45] does not mean that the social activity of mothering is read deterministically onto the biological fact of childbearing.

All the evidence is that human infants, with their plastic, adaptive brains and ready capacity to learn, develop social expectations concerning their own gender identity, and the activities appropriate to that gender, irrespective of their genetic sex and largely independent of any simple relationship to their own hormone levels (which can at any rate be themselves substantially modified in level by social expectations and anticipations). Psychocultural expectations profoundly shape a person's gender development in ways that do not reduce to body chemistry.

Claims for the Evolution of Patriarchy

The determinist argument does not stop however with merely reducing the present existence of the patriarchy to the inevitable consequence of hormonal balance and brain masculinization or feminization, but pushes insistently to explain its origins. For if the phenomenon exists, sociobiologists claim it must be adaptively advantageous and determined by our genes; it must therefore owe its present existence to selection for these genes during the early course of human history. Even if it were not now the case that patriarchy was the best of all thinkable societies, it must be the best of all possible societies because at some previous time in human history it must have conferred an advantage on those individuals who operated according to its precepts. This is the core of the Wilsonian thesis, just as it was of the earlier wave of pop ethology offered by, for instance, Tiger and Fox.[46]

In this thesis, the near universality of male dominance arose on the bases of the biological and social problems caused by the long period of dependency of the human infant on adult care, by comparison with other species, and of the primitive mode of obtaining food employed in early human and hominid societies—gathering and hunting. If a principal source of food was given by the hunting down of large mammals, which required long expeditions or considerable athletic prowess, even if men and women originally contributed equally to this task, women would be disadvantaged in such hunting by being pregnant or having to care for a baby they were breast-feeding, and indeed the baby's own life would be put in danger. So there would have been pressures on the men to improve their hunting skills, and on women to stay home and mind the children. Hence, genes that favored cooperative group activities and increased spatio-temporal coordination would be favored in men though not in women; genes that increased nurturative abilities—for instance, for linguistic and educational skills—would be favored in women. A socially imposed division of labor between the sexes became genetically fixed, and, as a result, today men are executives and women are secretaries.

It is easy to see the attractions of such evolutionary just-so stories, with their seductive mixture of biological and anthropological fact and fantasy. The existence of a sexual division of labor in primitive societies is a starting point as much for purely social accounts of the origins of patriarchy (for instance Engels)[47] as for biological. What is quite uncertain on the basis of newer anthropological evidence is the extent and importance of the hunter-gatherer distinction. In terms of overall contribution to food, gathering—a predominantly female activity—rather than hunting seems to have been the more important.[48] And in any event, with the small family sizes and spaced-out births of nomadic gatherer-hunter groups in the sparse conditions of their existence, the period over which women would have been at a physiological disadvantage in participation in hunting through being at a late stage of pregnancy or early stage of child-rearing would have been small.[49]

The point, however, is not to bandy about anthropological speculation, which can seemingly be directed to suit any case, but to emphasize that the real division of labor between men and women—which appears to have lasted, with variations and exceptions, over much of recorded history—still does not require a biological determinist explanation. Nothing is added to our understanding of the phenomenon, or of its persistence, by postulating genes "for" this or that aspect of social behavior. If patriarchy can take—in the Goldbergian sense—any form from baby-kissing to crusading, the leash on which genes hold culture[50] (whatever such a concept might mean) must be so long, so capable of being twisted and turned in any direction, that to speculate upon the genetic limits to the possible forms of relationship between men and women becomes scientifically and predictively useless; it can serve only an ideological interest.

From Animals to Human Beings and Back

The structure of the determinists' argument that we have discussed so far is as follows: contemporary society is patriarchal. This is the consequence of individual differences in abilities and proclivities between men and women. These individual differences are present from early childhood and are themselves determined by differences in the structure of male and female brains and the presence of male and female hormones. These differences are laid down genetically, and the genes for the differences have been selected as a result of the contingencies of human evolution. Each step in this reductionist argument is, as we have seen, fallacious or merely meretricious, a sort of magical hand-waving in the absence of data. Characteristically, however, the argument takes one final step—that of analogy with other species.

Again and again, in order to support their claims as to the inevitability of a given feature of the human social order, biological determinists seek to imply the universality of their claims. If male dominance exists in humans, it is because it exists also in baboons, in lions, in ducks, or whatever. The ethological literature is replete with accounts of "harem-keeping" by baboons, the male lion's domination of "his" pride, "gang-rape" in mallard ducks, "prostitution" in hummingbirds.

There are multiple problems associated with such arguments by analogy. Many derive from a common cause, the relationship between the subjective expectations of the observer and what is observed. We can consider three general areas of difficulty. First, inappropriate labeling of behavior. For instance, many species live in multi-female, single (or few) male groups, with excluded males living separately either isolated or in small bands. In such multi-female groups the male will tend to attack and drive out other conspecific males, denying them access to the females. The ethologists who observe this form of group living describe the female group as being the "harem" of the male. But the term "harem" defines a sexual power relationship between a man and a group of women that emerged in Moslem and some other societies at a particular time in human history. Harems were kept by princes, potentates, and wealthy merchants; they were the object of elaborate social arrangements; they relied heavily on the wealth of the male concerned; and they coexisted in the societies in which they occurred with many other forms of sexual relationships, including homosexuality and monogamy, if the literature of the period is to be relied upon. In what sense are the multi-female groupings among some species of deer or primate or among lions so to be regarded? Indeed, in the case of the lion grouping it is now clear that far from being "supported" by their male, it is the lionesses within the group who do the bulk of the hunting and provide most of the food.

An ethology that observes the nonhuman animal world through the lenses offered by its understanding of human society acts somewhat like Beatrix Potter; it projects, willy-nilly, human qualities onto animals and then sees such animal behavior as reinforcing its expectation of the naturalness of the human condition: Mothers are nurturative because Peter Rabbit's mother offers him camomile tea when he finally escapes being put into Mr. MacGregor's pie. In this way the behavior of nonhuman animals is persistently confused with that of humans. Inappropriate analogies make animal ethology harder to do. At the same time they form ideological refractions that seemingly reinforce the "naturalness" of the status quo in human societies.

A second problem area arises from the limited nature of the observer's account of what is happening in any social interaction. It is not merely that observed animal behaviors are inappropriately labeled; the observations themselves are partial. Studies of so-called dominance hierarchies tend to focus on a single parameter, perhaps access to food or who copulates with whom. Yet there is good evidence in several species that position along one dominance continuum—even if we accept the term—does not imply a matching position along other continua.

Studies of sexual behavior in animals are parlously distorted by the assumption, seemingly based on an almost Victorian prudery among ethologists, that the male is the main actor, heterosexual procreative sex is the only form to be considered, and the task of the female is merely to indicate willingness ("receptivity") and then lie down and think of England. Whether it is newts, ducks, or rats,[51] this androcentric fantasy works its way through the ethological literature. Only relatively recently has the female's role in courtship behavior ("proceptivity") become a more acceptable field of study, and it become recognized that, for instance, among rats it is mainly the female who initiates and paces sexual contacts.[52] It is surely no coincidence that the female's role in animal courtship has been discovered at a time when a new view of women's sexual independence has become current.

Thirdly, generalizations about the universality of particular patterns of behavior are made on the basis of data derived from small numbers of observations on a tiny number of species in a limited range of environments. It is well known that the study of primate ethology was seriously led astray over many years because the observations on which theories of aggressive intraspecific competition were built were made on populations in captivity in confined zoos, while the behavior of the same species in the wild was very different.[53] The same or closely related species of primates can live in widely differing habitats, varying for instance between mountains and savanna, and between conditions of relative abundance and relative scarcity of food. Under different circumstances, their social groupings and interrelations vary markedly. And as between the many different species—for example of primates —social and sexual groupings can vary from more or less monogamous to promiscuous, from groups with no recognizable dominance to those that seem more hierarchically ordered, from those that are male-led to those that are female-led, from those with marked sexual dimorphism to those with little sexual dimorphism.[54]

To select from this enormous abundance of animal observation only those moral tales that seem to support the naturalness of particular aspects of human sexual relationships and of the patriarchy is to imperil our understanding of both nonhuman and human social biology. If the tales selected by popular ethological accounts all appear to point in a single direction, one must ask: What interest is such a selective account serving? Just as understanding the behavior of baboons or lions is not helped by spurious analogizing from that of humans, so understanding the social biology of humans is not helped by reducing it to that of baboons.

These strictures remain regardless of who makes the reduction. It is not only defenders of the patriarchy who so unabashedly naturalize the arguments for innate differences in cognition, affectual understanding, and aggression between men and women. One school of feminist writing too has argued such an essentialist position, not merely stressing the importance of feminine, rather than masculine, ways of knowing and being, but rooting them in women's biology. This is the force of the defense of the right hemisphere offered by

Gina, to which we referred earlier, and it forms the basis of Firestone's argument in the *Dialectic of Sex,*[55] which sees, as does the strand of radical feminism that has followed her, the primary division in society as arising not from the division of labor by class and gender but from the biological differences between men and women.

A strand of feminist sociobiology has arisen that has centered on female rather than male evolutionary adaptation as the motor of social change during the transition from hominid to human societies. In part this focus on women has been a necessary redressing of the androcentric view offered by the dominant sociobiological strand; but to repeat the methodological errors of masculinist science in the process is to offer merely the other side of the same false coin.[56]

The essentialist argument echoes that powerful tradition in psychoanalysis which seeks the roots of differences in behavior between the sexes as lying, if not in the brain, then in the ineluctable biology of the genitalia. For the Freudian tradition it is the discovery by boys that they have a penis and by girls that they lack one that is at the heart of subsequent differences in their behaviors. But where for Freud and his followers this is the source of penis envy in girls, a feminist psychoanalytic approach offers instead the argument that it is the women's power to conceive that is central; that men, alienated from their seed at the moment of impregnation, mourn this loss thereafter and become committed to creating an external, object-centered universe of artifacts, a commitment that produces the overwhelmingly phallocentric culture of a male-dominated society.[57]

Removing the locus of male domination from the brain to the genitals and to the act of procreation does not, however, thereby avoid the methodological fallacies of seeking to reduce social phenomena to nothing but the sums of individual biological determinants, and of seeking simplistic "underlying" unitary explanations of diverse cultural and social phenomena. Where for Wilson genes hold culture on a leash, for the theorists of phallocentrism it is the penis and vagina that do so. Yet important though the male-female dialectic is, it cannot be the only, or indeed the underlying, determinant of the vast variety of human sexual and cultural forms. Not merely does such essentialism assert primacy over the struggles of class and race, but it lays claim to a universality that transcends both history and geography.

We must be more modest. We do not know the limits that biology sets to the forms of human nature, and we have no way of knowing. We cannot predict the inevitability of patriarchy, or capitalism, from the cellular structures of our brains, the composition of our circulating hormones, or the physiology of sexual reproduction. And it is this radical unpredictability that is at the core of our critique of biological determinism.

Subjectivity and Objectivity

There is one final point to be made. In this chapter we have tried to analyze the structure and fallacies of the biological determinist argument which, beginning from the indubitable fact of patriarchy in present-day industrial societies, seeks to ground that phenomenon in a biological inevitability. We have insisted that although all future as well as all past forms of relationship between men and women, both individually and within society as a whole, must be in accord with human biology, we have no way of deducing from the diversity of human history and anthropology or from human biology or from the study of ethology of nonhuman species the constraints, if any, that such a statement imposes.

What can be said, however, is this: We have described the emergence of biological determinist and reductionist thinking within science as an aspect of the development of bourgeois society over the period from the seventeenth century till the present day. This society, however, is not merely capitalist but also patriarchal. The science that has emerged is not merely in accord with capitalist ideology but with patriarchal ideology as well. It is a predominantly male science, from which women are squeezed out at all levels—excluded at school, frozen out within the university, and relegated to a reserve army of scientific labor, the technicians and research assistants to be hired and fired but not to be distracted from their main task of domestic labor, of nurturing the male scientist and rearing his children.[58] The story of how these exclusions operate has been told by women many times now.[59] Exclusion has a double effect. First, it denies half of humanity the right to participate equally in the scientific endeavor. Second, the residual "scientific endeavor" that the male half of humanity is left to practice on the back of the domestic and reproductive labor of women itself becomes one-sided.

It has long been recognized by historians of science that Greek science, where theory was divorced from practice, was a peculiarly patrician form of knowledge, precisely in that those who developed it were spared the day-to-day need of practice by the existence of a slave population that did the work. It was the unifying of theory and practice offered by the coming together of science and technology in the industrial revolution that generated the specific modern form of scientific knowledge. But just as Greek science was ignorant of practice and could not advance until that unity had been built, so today's patriarchal science is ignorant of domestic and reproductive labor and—as Hilary Rose has argued—is only and can only be a partial knowledge of the world.[60]

The particular stress that patriarchal science places on objectivity, rationality, and the understanding of nature through its domination is a consequence of the divorce that the division of productive and reproductive labor imposes between cognition and emotion, objectivity and subjectivity, reductionism and holism.[61] Such patriarchal knowledge can only be partial at best; feminist critiques of male-dominated science, by reemphasizing this scientifically neglected or rejected half of the interpretation and understanding of experience, are beginning to move from the analysis of reductionism to the creation of new knowledges.[62] In the long run only the integration of both forms of knowledge—just that integration which reductionism denies is necessary and determinism denies is possible—must be our goal.

CHAPTER SEVEN

ADJUSTING SOCIETY BY ADJUSTING THE MIND

The Politicization of Psychiatry

In the early 1970s, rumors of a wave of political dissidence among the Soviet intelligentsia—particularly scientists—began to reach the attentive ears of Western journalists. The dissidents were raising a variety of issues: their desire for greater freedom to travel and make contact with scientists abroad, their concern over the direction of Soviet internal and foreign policy, and what were later to become known as "human rights" issues. The response of the Soviet state to these challenges seemed only in the last instance to be one of direct political or administrative repression; more frequently individual protestors were harassed, brought in for psychiatric investigation, diagnosed as being mentally disturbed—typically, schizophrenic—and then incarcerated in psychiatric hospitals.[1] The paradigm case was that of biochemist Zhores Medvedev, who had written books discussing the weaknesses of Soviet science, the censorship system, and the Lysenko affair. In 1970 Medvedev was subjected to involuntary psychiatric examination, diagnosed as suffering from "schizophrenia without symptoms," and hospitalized. (Among the diagnostic signs for this schizophrenia, as Medvedev later described it in his book *A Question of Madness*[2] were "being interested in two things simultaneously, science and society.") Within the hospital, Medvedev was "threatened" with the use of psychotropic drugs, and only pressure from inside and outside the Soviet Union and the energetic intervention of his brother Roy resulted in his release after a few weeks and subsequent exile in England.

The protests concerning the plainly "political" use of psychiatry among concerned journalists and academics in the West were loud; there was much pressure on the World Psychiatric Association to censure Soviet psychiatry and boycott professional meetings organized by the Soviets.[3] The WPA finally passed an appropriate resolution in 1977, and the Soviet Union subsequently withdrew from the organization. The WPA's reluctance to take up a position, despite the fact that in the Medvedev and similar cases it was transparent that the role of psychiatry was to medicalize a political question and by doing so to depoliticize it, is interesting. It is important to see that it was not so much that the Soviet protesters were being punished for their protest, although they themselves clearly believed that they were. Rather, the state seemed concerned to *invalidate* a social and political protest by declaring the protesters to be *invalids*, sick, in need of care and protection to cure them of their delusions that there were any blemishes on the features of the Soviet state. But we would argue that the forensic and other psychiatrists who are asked to diagnose the disease of the Soviet protesters are not behaving very differently from their counterparts in the West. Perhaps the chief difference lies in that while the most common candidates for psychiatric hospitalization in the West are drawn from among the working class, women, or ethnic minority people who find it hard to locate a megaphone to speak their troubles to the waiting media world, the Soviet intelligentsia who have been hospitalized are neither inarticulate nor dispossessed.

This essential similarity perhaps accounts in part for the World Psychiatric Association's reluctance to take a political stand on the human rights question in the Soviet Union; there really is no major discontinuity between Soviet and Western practice. The clinical regimes and drug therapies offered in Russian hospitals are not, in the end, very different from those used in the West; the "threat" of being tranquilized or given a chemical straitjacket of chlorpromazine, so feared by the Soviet dissidents, is, as we shall show, part of the day-to-day experience of hospital and prison inmates in Western institutions. Psychiatrists in different countries are in substantial agreement on the symptoms to be regarded as diagnostic of schizophrenia; perhaps significantly, the widest criteria for schizophrenia of any of the nations studied came from the United States and the Soviet Union.[4] If we condemn the Russian psychiatrists as willing or merely compliant agents of political oppression, how may their Western colleagues avoid similar charges?

What, for example, are we to make of the use of psychiatric diagnoses in the treatment of young black offenders in Britain?[5] What do we make of the disclosures that in the late 1970s there were still women in British hospitals who had been confined since the 1930s for the "madness" of having illegitimate children?[6] Section 65 of the British Mental Health Act confines a patient for life in a secure hospital unless the home secretary permits that person's release or transfer. In 1980 Moss Side, a secure hospital in Manchester, contained a 21-year-old man confined for life; his "illness" (crime?) was that he had been caught stealing a trivial amount three years previously and had in anger smashed a jug and ashtray at his parent's home.[7]

There should be no misunderstanding; the point is not to "justify" the Soviet actions, which are as barbarous as those of any strong state that believes itself under threat, and which are certainly diametrically opposed to the human liberatory goals of socialism and communism. What we see in the action of the Soviet state however, is the mirror to the medicalized ideology of biological determinism in the advanced capitalist states of the West. Looking into that mirror enables us to see our own situation more clearly.

Especially over the last decade, biological determinist arguments have increasingly been heard insisting that the explanation for the symptoms of all social ills, from violence on the streets through the poor education of schoolchildren to the expressed feelings of the meaninglessness of life of middle-aged housewives, must be located in the brain dysfunction of the individual concerned. The first line of defense of the status quo is always ideology; if people believe that the existing social order, whatever its inequalities, is inevitable and right, they will not question it. In this way, as we saw in the context of IQ, ideas, ideologies, become a material force. In its declared aims of "correcting" the inappropriate thoughts of Soviet citizens, Soviet psychiatry is acting as an agent of ideological control.

It would be a mistake, however, to see the coercive use of psychiatry as merely a cynical attempt to suppress dissidents while pretending to help them, like the mystifying term "protective custody" introduced by fascist regimes in the 1930s to mean imprisonment or commitment to a concentration camp. The labeling of social dissidents as mad is but one aspect of a general attempt to understand and cope with social deviance. Despite the best efforts of the family, of peers, of the institutions of social indoctrination like the schools, the press, and the electronic media to produce correct thoughts and civilized behavior, some people persist in reaching the wrong conclusions and behaving badly. Such people are unreasonable and must be suffering from a brain defect or else they would see, as we do, how to think and act correctly. If, in addition, their thoughts and behavior threaten the very basis of society, the simple possibility of treating their madness medically becomes a social imperative. The medical model of deviance then provides even the most cynical state apparatus with a legitimate tool to control the behavior of individuals before they cohere as a dangerous social group. The last decades of medical and neurobiological research have generated a battery of technologies for the treatment, containment, and manipulation of dissident or abnormal individuals. The direct and immediate threats that these technologies pose are among the most disturbing with which this book has to deal. As we shall see, reductionist technologies are not disqualified from "working" merely because the ideology that frames them misaddresses the material world. Drugging people or cutting out parts of their brain will certainly change their behavior—may even make them less capable of protest—even though the theory on which the treatments are based is quite mistaken.

Violence and the Brain.

Soviet authorities strive to locate the social unrest that the individual reflects and participates in within the biological character of that individual. This same urge was well represented in the aftermath of the inner-city riots of the 1960s in the U.S. In a well-known letter to the *Journal of the American Medical Association*, three Harvard professors —Sweet, Mark, and Ervin—wrote on "The Role of Brain Disease in Riots and Urban Violence." Their argument was clear:

That poverty, unemployment, slum housing, and inadequate education underlie the nation's urban riots is well known, but the obviousness of these causes may have blinded us to the more subtle role of other possible factors, including brain dysfunction in the rioters who engaged in arson, sniping, and physical assault.

It is important to realize that only a small number of the millions of slum dwellers have taken part in the riots, and that only a subfraction of these rioters have indulged in arson, sniping, and assault. Yet, if slum conditions alone determined and initiated riots, why are the vast majority of slum dwellers able to resist the temptations of unrestrained violence? Is there something peculiar about the violent slum dweller that differentiates him from his peaceful neighbor?

There is evidence from several sources . . . that brain dysfunction related to focal lesion plays a significant role in the violent and assaultive behavior of thoroughly studied patients. Individuals with electroencephalographic abnormalities in the temporal region have been found to have a much greater frequency of behavioral abnormalities (such as poor impulse control, assertiveness, and psychosis) than is present in people with a normal brain wave pattern.[8]

Shortly afterwards, Mark and Ervin received substantial research grants from the U.S. Law Enforcement Assistance Agency, and the philosophy came to full flower in their book *Violence and the Brain*. The thesis was simple: Whatever the original causes of the brain dysfunction, its damage was deep and irreversible:

If environmental conditions are wrong at the important time, then the resulting anatomical maldevelopment *is irreversible*, even though the environmental conditions may later be corrected. . . .

The kind of violent behavior related to brain malfunction may have its origins in the environment, but once the brain structure has been permanently affected, the violent behavior can no longer be modified by manipulating psychological or social influences. Hoping to rehabilitate such a violent individual through psychotherapy or education, or to improve his character by sending him to jail or giving him love and understanding—all these methods are irrelevant and will not work. It is the brain malfunction itself that must be dealt with, and only if this is recognized is there any chance of changing behavior.[9]

Note that Mark and Ervin are not denying the existence of social problems in American society; they are endeavoring to protect "society" from the threatening responses of individuals to those problems. Violence, for them, is an inappropriate way for inner-city people to respond to the enforced poverty of their environment, to unemployment, or to racism, and it must therefore be eliminated. The brain's mechanisms of violence and aggression have gotten out of control; the proposed treatment is to find what the nineteenth-century natural philosophers knew as "the seat of passions," and to destroy it. A group of brain structures, the limbic system, is in some way involved in the passions of love, hate, anger, fear—what psychologists call "affect," because when any of the structures of this system are damaged or destroyed there is a permanent change in those aspects of personality. Reductionist neurobiology then identifies these structures as responsible for producing affect, and the surgical destruction of one of them, the amygdala, is Mark and Ervin's treatment for violence.

According to Mark and Ervin, as many as 5 percent of all Americans —11 million people—are suffering from "obvious brain disease," and an additional 5 million have brains that are "subtly damaged" so far as their limbic systems or affective responses are concerned. What is required is a mass screening program and an early-warning test to detect those individuals with low thresholds for violence. "Violence," they claim, "is a public health problem." The nature of this "problem" is perhaps better revealed in the following exchange of correspondence between the director of corrections, Human Relations Agency, Sacramento, and the director of hospitals and clinics, University of California Medical Center, in 1971.[10] The director of corrections asks for a clinical investigation of selected prison inmates "who have shown aggressive, destructive behavior, possibly as a result of severe neurological disease," to conduct "surgical and diagnostic procedures . . . to locate centers in the brain which may have been previously damaged and which could serve as the focus for episodes of violent behavior" for subsequent surgical removal.

An accompanying letter describes a possible candidate for such treatment, whose infractions while in prison include problems of "respect towards officials," "refusal to work," and "militancy"; he had to be transferred from one prison to another because of "his sophistication . . . he had to be warned several times . . . to cease his practicing and teaching Karate and judo. He was transferred . . . for increasing militancy, leadership ability and outspoken hatred for the white society . . . he was identified as one of several leaders in the work strike of April 1971. . . . Also evident at approximately the same time was an avalanche of revolutionary reading material." The director of hospitals and clinics replied to this request, agreeing to provide the treatment, including electrode implantation, "on a regular cost basis. At the present time this would amount to approximately $1000 per patient for a seven day stay."

Until public protest forced its disengagement, the Law Enforcement Assistance Agency proposed to fund the initial work of the California Center for the Reduction of Violence to the tune of some $750,000.[11] And such plans are not confined to the United States. In a similar vein, the West German authorities proposed a neuropsychiatric investigation of Ulricke Meinhof, one of the Red Army Faction militants arrested and imprisoned on charges of political violence, in the search for a biological "cause" for her political activity. Her death in prison preempted any final conclusions from this stab at medicalizing. The official British response to the inner-city riots of 1981 has so far avoided this approach, seeing no middle path between the reinforcement of ideological control—as with the recurrent emphasis by Margaret Thatcher and her successive home secretaries, Willie Whitelaw and Leon Brittan, on restoring the morality of the family and parental control over their children—and the full weight of an increasingly militarized police. It has been left to more liberal determinists to argue that the inner-city rioters may have an excess of lead in their bodies from gasoline fumes.[12]

Proposals for the direct surgical control of violence are but the tip of the iceberg of the coupled ideology and technology of behavior control as it has emerged in the last decade. It is true that the fantasies have outrun the reality. The chief science fiction visionary is perhaps Dr. Jose Delgado, whose 1971 book *Physical Control of the Mind: Towards a Psychocivilized Society*[13] set the agenda for the decade. Based on his experiments on direct implantation of stimulating and receiving electrodes into the brains of hospitalized patients and experimental animals, he claimed to be able to modify mood and behavior by the stimulation of appropriate limbic system sites. The electrodes can be monitored and impulses fed into them by remote control. In Delgado's hands the possibilities that this opened up in an era of microelectronics are that

it may be possible to compress the necessary circuitry for a small computer into a chip that is implantable subcutaneously. In this way, the new self-contained instrument could be devised capable of receiving, analyzing and sending back information to the brain, establishing artificial links between unrelated cerebral areas, functional feedbacks and programs of stimulation contingent on the appearance of predetermined wave patterns.[14]

What are the potentials opened up by such a system? According to the proselytizer of law enforcement by brain control, one possibility would be:

A transponder surveillance system can surround the criminal with a kind of externalized conscience—an electronic substitute for the social conditioning, group pressures, and inner motivation which most of the society lives with.[15]

And if the conscience didn't work too well, then:

It is not impossible to imagine that parolees will check in and be monitored by transmitters embedded in their flesh, reporting their whereabouts in code and automatically as they pass receiving stations (perhaps like fireboxes) systematically deployed over the country as part of one computer-monitored network. We may well reach the point where it will be permissible to allow some emotionally ill people the freedom of the streets providing they are effectively "defused" through chemical agents. The task, then, for the computer-linked sensors would be to telemeter, not their emotional states, but simply the sufficiency of concentration of the chemical agent to insure an acceptable emotional state. . . . I am not prepared to speculate whether such a situation would increase or decrease the personal freedom of the emotionally ill person.[16]

There may at first seem to be an inconsistency between the biological determinist claim that biologically determined traits are immutable and their program to cure, say, violence by a program of drug or surgical intervention. The question, however, is a practical rather than a theoretical one. Biological determinists, as reductionists, necessarily hold that any human mental characteristics could in principle be changed by appropriate physical intervention at the level of the individual nervous system or metabolism. In practice, however, they distinguish between characteristics of a small minority of persons who display "deviant" behavior, and traits that are continuously distributed over a range, like IQ, or are said to be human universals, as, for example, territoriality.

When a small number of people display a deviant, and presumably undesirable, trait, the reductionist program prescribes an alteration in the gene or genes that are thought to determine the trait. If a defective gene is the ultimate cause of a deviant behavior, then an alteration in that gene cures the deviance. Since, in fact, no one has ever actually located any gene or genes for criminal violence, schizophrenia, or paranoid delusions, the treatment offered is at the level of anatomy or biochemistry; that is, at the level of the primary effect of the supposed genes. Nevertheless, genetic manipulation remains the ultimate goal of reductionist determinism.

In the case of traits with a continuous range like intelligence, or that are a part of the claimed universal human nature, no intervention at the level of individuals is practical, even if it were deemed desirable. The prospect of changing these genes or operating on the brains of a large fraction of the world's population is absurd. Determinist theory, then, claims these traits to be immutable, not for any deep theoretical reason, but only as a consequence of the limitations of human time and effort.

CHEMICAL FIXES

We need not enter the realm of science fiction to see attempts to manipulate directly the behavior of those defined as criminal or deviant. Far more common than the use of brain chopping and shocking techniques is the attempt to get a chemical fix on behavior. The use of drugs to control institutionalized people, in prisons or hospitals, is of course widespread. In Britain, continental Europe, and the United States, the prisons have already become testing grounds for such methods. Male sexual offenders are regularly being treated with cyproterone acetate, which renders them impotent and is described as the chemical equivalent of castration.[17] The massive use of psychotropic drugs in prison, from minor tranquilizers to the chemical straitjacket of chlorpromazine (the term is that of the psychiatrists who employ it, not ours), has been described time and again by prisoners and exprisoners, despite official denials by, for instance, the Home Office in Britain.

Official figures of British prison use of drugs have become available as a result of public pressure and show a scale of use far in excess of what could reasonably be expected for therapeutic purposes. It is of interest that prescription of sedatives, tranquilizers, and other psychoactive drugs per head of the prison inmates is lower in prisons for the psychiatrically disturbed, like Grendon, than in some general prisons like Brixton and Holloway. In 1979 the average central nervous system drug dosage rate at Grendon was 11 per person per year; in Brixton 299; in Parkhurst 338; and in Holloway, a women's prison, an astronomic 941 doses per inmate per year.[18]

AVERSION THERAPY

The rationale behind the use of drugs is control of behavior; how much more effective it would be, then, to go even further and to control thoughts before behavior begins. "Aversion therapy," in which the individual is taught to associate criminal or deviant thoughts, or behavior disapproved of by the prison staff, with nausea, sickness, muscular paralysis, or terror induced by drugs such as anectine or apomorphine or even by electroshock treatment, has been practiced experimentally or possibly even routinely in a number of American prisons (for instance, Vacaville, California, and Patuxent, Maryland). There have been eloquent testimonies to the terrifying and brutalizing effects of such strategies.[19]

But the use of drugs to modify behavior inside institutions is only a symptom of the much wider search for chemical fixes outside, in the community at large. In Britain today no fewer than 53 million prescriptions for psychoactive drugs—about one per head of the population—are issued each year.[20] It is important to emphasize the scale of this use, and that the typical drug user in advanced capitalist societies is not a pot-smoking, alternative life-style teenager, or even a down-and-out alcoholic tramp, but a middle-aged housewife sustaining herself with her uppers and downers through the rituals of daily existence.

In a coercive and stressful society the individual has, baldly, two courses of action available: to struggle to change his or her social circumstances, or to adapt to the social conditions. The massive utilization of psychotropics is part of the mechanism of adjusting the individual to the status quo, of hyping, sedating, or tranquilizing the emotions. People trim or stretch themselves—or are trimmed or stretched by the medical authority of others—to fit the Procrustean bed of contemporary society, which insists on shaping its citizens into happy—or at least uncomplaining—consumers, if they are not to be expelled or institutionalized as congenitally unfit.

Let us once again emphasize that we are not denying that drugs work; of course they affect our emotions, thoughts, and behavior, in ways that we will return to below. And faced with unendurable pain, drugs offer one—sometimes the only—way of masking it. But they don't cure it. Given a toothache, aspirin will make life tolerable—but only till one can locate a dentist. The technology of drug control offers no dentists, only biologically determined causes in which the pains of existence are our fault, for not responding adequately to the challenges of our environment.

BEHAVIOR MODIFICATION

Aversion therapy seems the model of a biologistic method of achieving control over the behavior of another human. Yet its theory is derived explicitly from Skinnerian behaviorism—and earlier in this book we have described such behaviorist theories as representing a form of cultural determinism. Skinnerian psychology sees all human behavior as the consequence of past histories of "contingencies of reinforcement." The individual begins as a *tabula rasa* and learns to behave in particular ways as a result of rewards or punishment, subtly or less subtly administered by his or her "environment" or parents, teachers, and peers.[21] Even infant speech, according to Skinner, is learned in response to mechanical (albeit unconsciously given) parental rewards or disapprovals for the words the infant acquires.

Not all therapies offered by behaviorist theoreticians demand the use of chemistry. Drugs are after all but one way of achieving a negative reinforcement; it can also be produced by placing the individual into a controlled environment in which failure to conform to the desired behavior (for example docility or deference to prison guards) is punished by withdrawal of privileges, solitary confinement, restricted diet, and so forth, while "good" behavior is appropriately rewarded. If this does not sound particularly forbidding, it should be pointed out that the controlled environment may include "boxcar cells," as in Marion, Illinois, which are described by Samuel Chavkin as follows:

These are cubicles that are cut off from the rest of the penitentiary by two doors: a steel door to shut out the light and a covering plexiglass door to keep sound from coming in or out. A prisoner suddenly taken sick has no way of making it known, no matter how loudly he may be shouting for help. Ventilation is poor and one 60-watt bulb supplies the light. Fifty of the most outspoken inmates, some of whom were known to have communicated with their congressmen and news media protesting their plight, have been placed in the boxcars.[22]

According to one boxcar resident, Eddie Sanchez:

It has been very hard not to lose hope. To tell the truth I've just about lost hope. I feel I will be killed by my keepers. I really don't fear death. I've faced it often before. I do have one regret and that is that I've never been free. If I could be free for one week, I would be ready to die the next. Is it any wonder I don't believe in God? I can't picture a God as cruel as to deny a person even a passing memory of freedom."[23]

More dramatic treatments are at use in Patuxent, Maryland, where, again according to Chavkin:

Treatment of "defective delinquents" employs a "restraining sheet" for noncooperative inmates. As described by a reporter in the Washington *Daily News*, this "is a device in which a naked inmate is strapped down on a board. His wrists and ankles are cuffed to the board and his head is rigidly held in place by a strap around the neck and a helmet on his head. One inmate testified he was left in the darkened cell, unable to remove his body wastes. He said he was visited only when a meal was brought. Then, one wrist was unlocked so he could feel around in the dark for his food and attempt to pour liquid down his throat without being able to lift his head."

An additional terror tactic used at Patuxent is that of holding the prisoner for an indefinite sentence, his liberation being dependent on the psychiatrist's prognosis as to the inmate's dangerousness in the future.[24]

We have no such clear description of the nature of treatments offered in the Behavior Control Units of British prisons; they certainly have involved at various times sensory deprivation, restricted diet, solitary confinement, and loss of remission.[25]

Behavior modification theories seems to be used increasingly in the British school system. "Special units" (or "sin-bins") for behaviorally disturbed children (sometimes labeled as educationally subnormal—ESN) are frequent in some districts—for example, the London borough of Haringey.[26] Children considered "disruptive" in normal classes are then subjected to a specific regime of rewards and punishments, "token economies" by which they gain points for approved behavior that may be accumulated toward earning such privileges as being allowed out of the school for a period.

Behavior modification begins with a culturally determinist theory; in practice—at least in the experience of those it treats, whatever the declared intentions of its advocates—it is hard to see where it may be distinguished from the most explicit of biological determinist therapeutic programs. Both are essentially "victim-blaming," locating the problem inside the individual, who must be tailored to fit the social order that he or she so evidently mismatches at present. Both are the reverse of that slogan of 1968: "Do not adjust your mind; the fault is in reality." The clarity of that fault in reality becomes the more apparent when we hear that the Haringey sin-bins are disproportionately filled with young male blacks.

The paradox of cultural determinism generating a biological determinist therapy is only apparent, however. Both determinisms are reductionist, as we have explained earlier, and are related as the opposite sides of the same coin. For the liberal-minded biological determinist seeking to escape from the arid, inexorable rigidity of the view of human nature to which the theory has led, the escape is a sort of cultural dualism, which assigns a constraining effect to the genes, but allows a "long leash" to the individual personality. One sees this happening time and again in the writings of sociobiologists like Wilson, Dawkins, or Barash (see Chapters 9 and 10). However, as both forms of reductionism begin in theory by giving the individual ontological primacy over the social formation of which he or she is a part, both end in practice by endeavoring to manipulate that individual. Because, irrespective of theory, biological methods of manipulation, by drugs or electroshock, are apparently much more powerful than the less direct methods of brain manipulation offered by the talking therapies, they must inevitably come to the fore when therapists or controllers are under pressure for quick solutions. Nowhere is this more apparent than in the quick slide from the definition of the "behavioral" category of "hyperactivity" to the organic diagnosis of minimal brain dysfunction to which we now turn.

Minimal Brain Dysfunction

The British classify troublesome youngsters as naughty, disturbed, or ESN, and place them in special schools; the "cause" is defective socialization—for example the lack of parental control, or adequate male role models in black families. For the United States, such deviant behavior in the young became a disease during the 1960s. The victims were boys some nine times more often than girls. The afflicted children were overactive in the classroom, they often interrupted the teacher, they did not tolerate frustration well, and they did not concentrate well. Though they seemed to be bright enough, they did not master their school subjects. The parents of these children, when questioned, often agreed that they were difficult children at home. This unhappy state of affairs was not the fault of the school system, or of the family, or of the larger society. It was a disease, the "hyperactive child syndrome." The problem was that the children had biologically defective brains. The defects were small and subtle, and could not be seen under even the finest microscope. Thus the term "minimal brain damage," soon to be replaced by "minimal brain dysfunction" (MBD), came into common usage.

The U.S. Department of Health, Education and Welfare defined MBD as referring to children of near average, average, or above average general intelligence with certain learning or behavioral disabilities . . . associated with deviations of function of the central nervous system. These deviations may manifest themselves by various combinations of impairment in perception, conceptualization, language, memory, and control of attention, impulse or motor function. . . . During the school years, a variety of learning disabilities is the most prominent manifestation.[27]

These problems were defined as biological and medical in nature. Thus, reasonably enough, the proposed remedy was to treat the offending children with drugs.

Within a couple of years many hundreds of thousands of American schoolchildren (as many as 600,000, according to some estimates) labeled as MBD, hyperactive, or learning disabled were receiving regular doses of stimulant drugs. The allegedly favorable response of the overactive children to stimulant drugs was said to be "paradoxical." With virtually no understanding of how such drugs might operate, or what their long-term effects might be, they were heavily and successfully advertised by the drug companies for the treatment of problem children. Wender,[28] in an influential book on MBD, urged that *all* children diagnosed as hyperactive should first be treated with drugs. Then, for those few who did not respond properly, other forms of treatment might be considered. The physician who failed to drug a hyperactive child was, in Wender's view, guilty of medical malpractice. The number of hyperactive children at large in the community had not been accurately determined—but they were legion.

When such children were intensively studied for evidence of neurological damage by Werry and his colleagues no "hard" signs could be found.[29] There were, however, numerous "soft" signs—difficult to elicit and hard to quantify—which might suggest the officially declared "deviations of function of the central nervous system." The soft signs included such things as general clumsiness, poor coordination, confusion between left and right, and the FLK syndrome—being a "funny-looking kid." Werry et al. were convinced that hyperactivity in otherwise normal children was "organic." That did not mean, however, that environmental factors played no role. They suggested that hyperactivity was "a biological variant made manifest by the affluent society's insistence on universal literacy." These biologically different children would have gotten along just fine if we had not insisted on trying to educate them!

The view that MBD and hyperactivity manifest themselves primarily in the classroom is widespread. Thus, books written for practicing physicians stress that the hyperactive child may be as docile as a lamb in the doctor's office. The organic impulse toward uncontrollable activity expresses itself only in the "structured task situations" of school and home. Thus the physician should not hesitate to prescribe drugs for a child described as overactive by its teachers or parents—even if the physician has not observed any overactivity. The very specific connection between "organic" hyperactivity and the schoolroom is remarkable.

Weiss et al. followed up groups of hyperactive and control children into young adulthood.[30] Questionnaires were sent both to their last high school teachers and to their present employers. The questionnaires dealt with whether the subject completed his assigned tasks, whether he got along well with peers and authorities, whether he could work independently, whether he would be welcomed back to the school or job, etc. The teachers rated the "hyperactives" as worse than the controls on every measure, to a significant degree. The employers drew no such distinctions; what differences occurred tended to favor the hyperactives.

There is not much to object to in the opening sentences of Roger Freeman's book *The Hyperactive Child and Stimulant Drugs*. With admirable candor, Freeman wrote:

There is only one phrase for the state of the art and practice in the field of minimal brain dysfunction (MBD), hyperactivity (HA), and learning disability (LD) in children: a mess. There is no more polite term which would be realistic. The area is characterized by rarely challenged myths, ill-defined boundaries, and a strangely seductive attractiveness.[31]

Though it is sometimes not considered polite to mention such matters, it is possible that financial profit has something to do with the seductive attractiveness of the area. There are enormous sums of money involved in the development and marketing of prescription drugs for troublesome children. The drug companies have not been hesitant to underwrite the research of scientists working in the area.

There is also good reason to believe that many of the stimulant drugs supposedly manufactured and prescribed for children find their way, at enormously inflated prices, to the illegal drug market.[32] The most commonly prescribed drug for hyperactive children, an amphetamine-like substance, is Ritalin (methylphenidate). By 1973 Omenn noted:

Illicit traffic in Ritalin has increased among narcotic addicts. . . . Those on Methadone appreciate the "up" effect of Ritalin. Those on heroin can prolong the duration of action of a given dose of heroin by concomitantly taking Ritalin. . . . In Chicago's Cook County Prison, Ritalin is called "West Coast" by the heroin addicts.[33]

Though the use of Ritalin and other stimulant drugs with hyperactive children is now common in the U.S., there is still astonishingly little evidence to show that the drugs produce any genuinely helpful effects.[34] There are great technical difficulties in evaluating whether a drug has any effect on behavior over and above the well-known "placebo effect." To deduce whether anything more than the power of suggestion is involved, it is necessary that both the observer and the child be "blind"; that is, they must not know whether the child has received the actual drug or an inert substitute for it. The stimulant drugs, however, tend to have powerful side effects—insomnia, loss of weight, fearfulness, depression—so both the child and observer can often detect when a placebo has been substituted for the real drug. To complicate matters even more, the behavioral changes said to be brought about by the drugs are difficult to measure. The studies thus often depend upon subjective ratings of the child's behavior made by the teacher or parent. There is no wonder that the drug literature contains a thicket of fragmentary and contradictory research results.

There is some indication, however, that at least over the short term Ritalin may cause children to squirm less in their seats in school and perhaps to pay more attention to some experimental tasks administered by psychologists. The positive results obtained in short-term studies were broadcast widely and helped to create a climate in which drug treatment was easily accepted. The common side effects of the drug were less often cited. There was much wonder expressed at the paradoxical effect of a stimulant drug on hyperactive children—it appeared somehow to calm them. The paradox has since disappeared: It is now known that the measurable effects of so-called stimulant drugs are similar in hyperactive and in normal children.[35] As to why stimulant drugs should be calming in children at all, this more general paradox is itself derived from a psychiatric and neurobiological naiveté that believes in a single site and mode of action for any drug—a point to which we return in the next chapter. Only the most socially isolated of nondrinkers would believe that a double whiskey always had the same effect on the person taking it.

There is no evidence whatsoever that the long-term administration of Ritalin has any beneficial effect on the symptoms and problems that cause children to be labeled as MBD or hyperactive. Weiss et al. looked at hyperactive children who had been treated with Ritalin for up to five years, and compared them to similarly hyperactive children who had not had drug treatment.[36] The long-term nature of this study is unprecedented in the Ritalin literature. The authors fully expected to observe a beneficial drug effect, and had prescribed the drug in their own clinic. They reported no differences in adolescence between the drugged and the undrugged children in school marks, in number of grades failed, in amount of hyperactivity, or in antisocial behavior. The problems of organically hyperactive children seemed to linger on, whether or not they had been drugged.

The most recent review of drug effects by Cantwell asserts that Ritalin "produced an improvement rate of 77% in hyperactive children.[37] What, however, is meant by "improvement"? The answer given by Cantwell is "a consistent positive effect on behavior which is perceived by teachers as disruptive and socially inappropriate." The alleged improvements, as Cantwell indicates, are not always easy to describe. Does the drug reduce excessive motor activity? That depends upon whether one measures "ankle movements or seat movements" and also depends upon "the situation in which the activity is being measured . . . on laboratory tasks . . . stimulants consistently decrease the activity level . . . in the playground . . . children . . . actually have an increase in activity level."[38] The picture of an organic brain dysfunction producing fidgety seats but quiet ankles, boisterous schoolroom behavior and inhibited playground behavior, is not entirely convincing. The organic basis of hyperactivity—and the continued prescription of drugs for untold numbers of children—clearly requires some bolstering.

The "Genetics" of Hyperactivity

There has been much effort devoted to the attempt to demonstrate a genetic basis for the hyperactive child syndrome. The peculiar logic of biological determinism suggests that implicating the genes in the "disorder" would justify treating it with drugs. To demonstrate the role of the genes, the first prerequisite, as always, is to show that the disorder runs in families. That was supposedly done in a study by Morrison and Stewart.[39] Those authors began with fifty children (forty-eight of them boys) who had been diagnosed as hyperactive in a hospital outpatient department. There were fifty control subjects, matched to the hyperactives for sex and age, who had been admitted to the same hospital for surgery. The parents of all children were interviewed, and they were questioned about other members of their families. The interviewer knew which children were which but was said to have conducted the interview "with no hypothesis in mind." The supposed *control* families included nine children (18 percent) said by the parent to be "hyperactive, wild, or reckless . . . or whose parents had sought professional help." Those nine cases were transferred from the control to the hyperactive group. Then it was discovered that a number of nasty disorders were significantly more frequent among the parents of hyperactives than among the parents of the now shrunken number of controls. The disorders more frequent among the parents of hyperactives were alcoholism, "sociopathy," and "hysteria." From the comments made by parents during the interviews, the authors— who also knew which children were which—felt able to make retrospective diagnoses of whether the parents had themselves been hyperactive children. The authors thought that more parents, aunts, and uncles of the hyperactive subjects had themselves been hyperactive children. Peculiarly, there was no report made on whether hyperactivity was more frequent among the siblings of the hyperactives than among those of the controls. When reporting an earlier study, Stewart et al. had indicated that 16 percent of hyperactive subjects—and 25 percent of control subjects—had hyperactive sibs.[40]

This study, according to the authors, was done "to see if we could find evidence that this behavior pattern was inherited." The results, again according to the authors, "suggest that the 'hyperactive child

syndrome' passes from one generation to another," and "the prevalence of alcoholism . . . favors a genetic hypothesis." The authors noted
that their findings were consistent with a 1902 report that "disorders
of intellect, epilepsy, or moral degeneracy" were common in the families of hyperactive children. The present appeal to the concepts of
degeneracy and poor genetic stock, however, was printed in 1971, in
a journal entitled *Biological Psychiatry*. The cases of "alcoholism"
among parents, it might be noted, were almost all males; all cases of
"sociopathy" were males, and all cases of "hysteria" were females. The
forms in which inherited bad blood erupted are evidently different for
the two sexes—but blood does tell.

The findings of Morrison and Stewart were soon repeated by Cantwell, who studied fifty boys who had been diagnosed as hyperactive
in a Marine Corps dependents clinic.[41] The control boys, drawn from
the pediatric clinic at the same marine base, were matched to the
hyperactives for age and social class. They had been screened in advance "to assure that there was no hyperactivity in their family." The
parents of all subjects were interviewed, and the results duplicated
those reported above. There was much more alcoholism, sociopathy,
and hysteria diagnosed among the parents of the hyperactives than
among those of the controls. From interviewing parents, the investigator was also able to diagnose alcoholism, sociopathy, and hysteria in
the grandparents, aunts, and uncles. There was more such degeneracy
among the relatives of the hyperactives. From the interviews, retrospective diagnoses of hyperactivity were also made—for parents, aunts,
uncles, and cousins. There was said to be more hyperactivity among
the relatives of the hyperactives. These data, based upon such long-
distance diagnoses, appeared in a scientific journal published by the
American Medical Association. They were subjected to elaborate statistical tests, with a clear belief that they were scientific. The author
pointed out that studies of the wives and relatives of male felons had
also found high rates of alcoholism, sociopathy, and hysteria; and he
concluded that "the hyperactive child syndrome is passed from generation to generation."

The same set of authors, having established to their own satisfaction
that hyperactivity runs in families, now attempted to separate genetic
from environmental factors by studying adopted children. Thus Morrison and Stewart contrasted thirty-five adopted children, diagnosed as

hyperactive, to the hyperactive and control children of their 1971 study.[42] The adoptive parents of the hyperactive adoptees, like the biological parents of the control children, were said to display no sociopathy or hysteria, and very little alcoholism. They were thus better quality people than the biological parents of the hyperactives reported on in 1971; there was little pathology in their families, and few retrospective diagnoses of hyperactivity. There was no information available about the biological parents, and families, of the hyperactive adopted children. The results meant, according to the authors, that "a purely environmental hypothesis of transmission for this condition cannot be sustained." That is, since Morrison and Stewart did not diagnose pathology in the adoptive parents of children who became hyperactive, the children must have been made hyperactive by their genes. Parents who are allowed to adopt, of course, have been carefully screened for the absence of pathology, so it is not surprising that Morrison and Stewart detected little pathology among them. We do not know whether hyperactivity is more (or less) common among adoptees than among ordinary children. The design of Morrison and Stewart's adoption study was repeated by Cantwell, who reported very similar results.[43]

There is a curious omission from the adoption studies. The studied children also had siblings, and in the case of the adopted children, step-siblings. The incidence of treatment for hyperactivity among the siblings would be of considerable interest; for example, do the biological children of the adoptive parents of hyperactive adoptees have a high rate of hyperactivity? If so, it would implicate family environment; but the easily obtainable sibling data were not presented.

Blaming the Child

There is a recurring theme in the literature on hyperactivity pointing out that those who have not had to deal with such children cannot appreciate how truly disruptive they are. The classroom is said to be thrown into turmoil by a hyperactive child, and teachers are driven out of their wits. Thus, even if stimulant drugs do not benefit the hyperactive child, they may at least quiet him or her enough so that others in the classroom can learn. This may be construed as an ingenious rationale for the continued use of a drug that does not help the drugged child. However, there is no evidence to show that classmates of a drugged hyperactive child learn more, or benefit in any other way, as a result.

Mash and Dalby have urged a greater "social system emphasis" in research, one which would focus on "the interaction between hyperactive children and their parents, teachers, peers, and siblings.... Little attention has been given to the effect of the hyperactive child in his social system."[44] Campbell et al., for example, reported that teachers tend to be more negative toward nonhyperactive children when a hyperactive disruptive child is in the class. That is, "hyperactive disruptive" children evidently turn teachers into monsters, who then act negatively toward all the children in their class.[45]

This important research finding by Campbell et al. was also cited by Helper in the *Handbook of Minimal Brain Dysfunctions*:

This carefully designed study also found evidence that the presence of a hyperactive child in a classroom affected interactions between the teachers and other children in that class. Teachers criticized the classroom control child in the classes containing a hyperactive child more frequently than the classroom control child in the classes of the nonhyperactive children being followed longitudinally in the study.[46]

The Campbell et al. report is actually a follow-up study of a group of children first described by Schleifer et al.[47] The study had begun with 28 hyperactive preschool children and 26 control children. As always, there was some loss of subjects between the time of the original study and the time of the follow-up; only 15 hyperactives and 16 controls were available for follow-up three years later. They were observed in their classrooms. Within each classroom the observer also watched a newly selected "classroom control" child of the same sex as the child from the original study. The children were observed in the classroom setting for one-half hour each, and all instances of "negative feedback" from the teacher were noted. That meant "expression of disapproval to child about the behavior or performance; reprimands." The observers were blind as to group membership of the children. This design meant that 31 separate classrooms were visited—15 containing a hyperactive child and 16 containing a control child from the original study. Within each of the 31 classrooms, a "classroom control" child was also observed. There was more negative feedback given by the teachers in the 15 classrooms containing a hyperactive child, but the negative feedback was given as often to the hyperactive child as to the classroom control. To be exact, the hyperactive children were observed to receive negative feedback an average of 0.67 times each—less than once during the half hour. Their classroom controls received negative feedback an average of 0.80 times each. The control children from the original study received negative feedback an average of 0.13 times, as did their classroom controls.

This rather modest-sized effect, even if taken at face value, is open to many different interpretations. To begin with, Campbell et al. had earlier reported that the Wechsler IQs of the control children in the original study were significantly higher than those of the hyperactives. The controls also came from a perceptibly higher social class level. The controls and the hyperactives were, at the time of follow-up, in different classrooms, with different teachers. Is it possible that teachers in schools attended by somewhat lower social class children, with lower IQs, behave more negatively toward their pupils?

We should not, however, take the Campbell et al. numbers too seriously. They claimed to observe a statistically significant effect, but this depended upon their inappropriate use of a statistical technique. They reported in addition (p. 241) three separate "t-tests"—a standard statistical technique. From tabled means and standard deviations presented on the same page it can be calculated that all three reported t-values are incorrect. The first report of the study (Schleifer et al.) indicated that there were only three girls in the hyperactive and three girls in the control group. When 41 of the original 54 subjects were followed up two years later (Campbell et al.) they included five hyperactive and two control girls. To find hyperactive-disruptive children guilty as charged, a more consistent and credible set of numbers than these should be required. The incorrect and contradictory statistics, however, appear in major scientific journals. They are solemnly cited as examples of social-system research, and they make their way into authoritative handbooks. The scientists attribute minimal brain dysfunction to the children. The children might with as much justice attribute it to the scientists.

Does Biological Determinism
Make for Good Therapy?

The drive toward the development of biological determinist theories for all aspects of the social condition and drugs for all disorders is powerful. Not all of it is to be understood simply in terms of the need to control and pacify an unruly population of prisoners, schoolchildren, and hospitalized and general-practice patients. This is part of the story, but it is clearly not all.

The feeling of meaninglessness and alienation experienced by a substantial proportion of the American and European population are real, not myths. The pressure for solutions is therefore also real, and to some degree we all believe in the promise offered by modern medicine of chemical solutions, whether we are patients or doctors. The demands by sufferers for an easing of their psychic pain, and the search by sympathetic doctors for such solutions, are powerful motors. The increasing prestige of molecular biology, with its apparent determinist certainties, offers the theoretical lure. The practical spur is the need for drug companies to circumvent the patent regulations by generating alternative formulations or mildly different chemicals by the assiduous labors of their organic chemists playing endless molecular roulette. According to World Health Organization figures, over 60,000 brands of drugs and other medicines are sold in the United States today, yet only 220 of these are considered necessary, well-documented drugs for well-documented disorders. Hence what the medical, psychiatric, and other caring professional services offer is a mix of therapies based on their—and their clients'—belief that something must be done; that the task of changing the social order is more daunting than that of fitting their clients to it. The therapeutic mix is determined only in part by the theory of the therapist; pressures of time and the blandishments of the drug companies are as important. Yet, taken as a whole, the pattern that results has all the qualities of the determinist, reductive argument described in the early chapters of this book. The point is that, here and now, biological determinists are in the actual business of proposing interventive strategies, drugs, neurosurgery, or behavioral therapies to control and modify human actions. One can well be forgiven for urging that medical or social interventions cannot wait until we get our theories right. Something must be done now. The question is not whether the explanations work but whether the treatment works. We are clearly not arguing that drugs or surgery are without effect on the behavior of the individuals to whom they are administered. Far from it; by our very definition of the ontological unity of human experience and action with human biology, if we administer drugs or cut circuits in the brain, the state of that brain will be changed, and this changed brain state will correspond to a change in behavior, experience, and action.

The issue is the relevance of such interventions to the diagnoses of the changes they are ostensibly designed to produce. Clearly, one way of preventing any individual from participating further in inner-city riots would be to sever the spinal cord at the neck, thereby separating the brain from the rest of the body and effectively preventing its subsequent functioning, an operation that can readily be performed even by relatively unskilled surgeons. Severing the spinal cord lower down is unlikely to be so effective, as it was reliably reported that individuals confined to wheelchairs took part in the inner-city riots and looting in Britain in 1980 and 1981. Similarly, inattention in class could be treated by drugs such as cyanide that block glucose oxidation in the brain, or those that interfere with nervous system transmitter functions, such as curare. These rapidly produce terminal effects in the individuals treated, who are therefore removed from the need to occupy the teachers' attention. Observation of these treatments by others who might be tempted into wayward activities could also produce a beneficial effect on their own brain chemistry, thereby preventing the spread of the disorder—a point emphasized as long ago as the eighteenth century, when the British Admiralty executed an admiral who had lost a battle at sea, in order, as Voltaire observed, "to encourage the others."

We are not being flippant. The essence of reductionist explanation is the assumption that a disorder is caused by a single simple malfunction of a body region, or a biochemical substance, or a gene. The notion of treatment by a "magic bullet," a specific drug intervention that has defined and unitary effects, is integral to one whole strand of medical thinking, typified by the claims that the causes of, for instance, smallpox or typhoid, are specific microorganisms. The treatments for the diseases are then programs of vaccination or immunization or antibiotic therapy. This is in contrast to the determinist view of low IQ, which is presumably the result of so many bad genes that no single magic bullet (only a magic sperm or ovum) will help. To explore these arguments in the sphere of general medicine will take us far afield; suffice it to say that epidemiological research makes clear that the concepts of disease causation and cure are far more complex than a simple germ theory or its equivalent. Whether or not particular microorganisms or viruses infect and produce disease in particular individuals in any given society is not straightforwardly predictable; for instance, the decline of cholera and tuberculosis over the past century is more attributable to general social and economic changes than to specific medical interventions at the level of the individual.[48]

We can illustrate this complexity and the inadequacy of the magic-bullet theory for brain and behavior, however. Take for example the arguments about violence and the brain, which claim that behavior can be changed by the removal of a particular brain region or the implantation of a set of stimulating electrodes. There is no doubt that taking out parts of the brain has some effects and that these are partially predictable. But brain lesions, whether the results of operations or accidents in humans, or in controlled animal experiments, have continued to produce puzzles and paradoxes.[49] In some areas relatively large volumes of brain can be "disconnected" without much obvious consequence—the huge regions of the frontal portions of the brain chopped out by psychosurgeons doing prefrontal lobotomy or leucotomy, for instance; in other cases very minute lesions have devastating effects, as when damage to a few cubic millimeters of tissue in the hypothalamus can profoundly affect an animal's eating, drinking, or sexual activities. Whether the lesions occur in youth or age, and the conditions under which recuperation and rehabilitation take place, will all profoundly affect the results.

All psychosurgeons know this,[50] even if they are rarely prepared to say so as crudely as the British doctor who described the case for psychosurgery in a woman who was a "compulsive" housecleaner and who spent all day washing, cleaning, tidying, and retidying her home, and became very depressed about it. An operation was performed and the patient was rehabilitated. And the result? Success, it would seem; the housecleaning stopped for the time being. But the woman was soon back at her compulsive cleaning practice, just as before, with one difference; now, instead of being depressed as she cleaned, she was quite cheerful about it. The surge of psychosurgery in the 1940s and 1950s, its relative decline in the 1960s, and its subsequent rebirth in more sophisticated form in the 1970s have been well told.[51] The point we wish to make here is that the fallacy that underlies the treatment is not merely one of reducing the social to the biological but of reducing the richness of the biological phenomena themselves.

The human brain consists of some hundred thousand million nerve cells, connected by the astronomic number of 10^{14} (one hundred million million) pathways. Like any successfully designed human-made machine, but in almost unimaginably more complexity, it is a system with built-in checks and balances and controls; the multiplicity of redundant pathways means that if any part of the system fails, or is damaged, as by psychosurgery, other parts will tend to take over the function that is lost. The result is that the consequences of operations or disease are either almost imperceptibly small and soon masked or they are so gross as to permanently impair the individual. Psychosurgery is almost bound either to be ineffective or to reduce the individual to a cabbage (and there has been no lack of critics of the use of psychosurgery in hospital practice to argue that this after all is one of its intended functions—it makes control easier for the hospital staff).

Psychosurgery is not much more precise than the work of a saboteur pulling out at random printed circuit boards from a computer. If you remove a transistor from a radio and the result is that the radio thereafter emits nothing but howls, you are not entitled to assume that the function of the transistor that was removed was that of a howl suppressor. Rather, what we see in the radio with the transistor removed is the workings of the rest of the system in the absence of the transistor. But the most likely effect of removing a transistor or disconnecting parts of the brain is indeed some sort of howl. Fortunately, brains are not merely vastly more complex than transistor radios, they also have considerable plastic capacity for regeneration or relearning. It is this fact that makes a very large proportion of even the laborious experimental work of the last fifty years on lesions in the brains of laboratory animals of limited theoretical value. This limit is even sharper for operations based on simplistic theories like those of Mark, Ervin, and their less media-conscious counterparts on human patients.

If psychosurgery's reductionism is so crude that it is incapable of achieving the stated purposes of its proponents, how about drugs? There is a similar but rather more complicated point to be made in this context. The interaction of any chemical, such as a drug, with the hundreds of thousands of different chemicals organized in rather precisely ordered spatial domains that constitute the biochemical map of the brain at any time are complex. The interactions vary from individual to individual and in any one individual at any different time. Think, for instance, of the many different ways in which any one of us might feel or act after taking orally dilute solutions of ethyl alcohol plus assorted aromatic esters. The organic substances present in wine, beer, or spirits readily enter the bloodstream and can be measured. Police tests for drunkenness in Britain are posited on the assumption that individuals with more than 80 mg per 100 ml alcohol in their bloodstream are too drunk to drive—but most people will know that a variety of different moods can be associated with such an alcohol level.

In an experiment described by the psychopharmacologist C. R. B. Joyce, two groups, each of ten individuals, are placed in separate rooms. In one room nine are given a "sedating dose" of barbiturates, one an "elevating dose" of amphetamine; in the second room nine receive amphetamine and the tenth barbiturates. In each room the odd individual out, rather than behaving in a way appropriate to the drug taken, behaves like the majority, either sedated on the amphetamine or high on the barbiturates. Not merely the extent but also the direction in which intake of a drug may change a person's mood, behavior, and so forth depends materially upon the social context. Indeed, merely telling a person that he or she has been given a drug that will alter mood, ease pain or depression, or whatever, is enough in a large number of cases to result in the individual reporting improvement. The placebo affect, as it is called, is very well known in clinical trials of psychoactive drugs. Thirty percent or more of individuals given "drugs" as treatment for depression report effects even if the drugs are made of biologically inert substances.

Of course, give enough of any drug and the result eventually becomes more easily predictable. Enough alcohol and the consequences are, crudely, stupor or death. And it is not without interest in the context of the supposed therapeutic use of Ritalin for MBD that it has been shown that whereas, on average, lower dosages may increase a child's attentiveness and "set" for learning, higher doses simply result in sedation—yet in school use it is the higher doses that tend to be employed.[52] This makes the drug yet one more version of the chemical straitjacket, ensuring that the teacher has an easier task in maintaining classroom order, but only by doping out the children who would otherwise make it harder.

There is a widespread medical belief that good drugs are like magic bullets that hit a single, precise target disease site (which may be a particular body tissue or a particular biochemical system). No drugs actually work like this, though; they can have a very wide range of effects both on biochemistry and behavior. Medical doctors and pharmacologists sometimes describe these as side effects—the very term is redolent of reductionist disappointment. Most of the interactions of extraneous drugs with the body's chemistry are more like an explosion with shrapnel flying in many directions and a large area of fallout rather than bullets producing a neat contained hole.

We can draw an example from the treatment of one "straightforward" disorder, Parkinson's disease. Sufferers from the disease show a characteristic tremor and shaking of the limbs—especially the hands—which becomes quite troublesome, for instance, when they have to pick up a cup or try to convey liquid to their mouths. The tremor is a result of the loss of control of fine motor movements. The nervous pathways that malfunction in Parkinsonism are known, and one of the chemicals involved in the transmission of nervous information down these pathways is a substance called dopamine. Hence the development of a drug called L-dopa, which interacts with the brain's normal dopamine metabolism and has been found to give some relief from Parkinson's symptoms. For a time L-dopa was seen as almost an archetype of a single-disease, single-cause, single-treatment therapy. Then it began to become clear that people treated with L-dopa were experiencing other things than merely relief of Parkinsonian tremor. Not only did the dose of drug have to be continually adjusted, but the individuals treated with it began to experience changes in their own existential state, with varieties of feelings of despair, elevation, entry into "hell" and hallucinations, as well as "organic" nervous system changes.[53] The drug, it turned out, interacted with many different systems within the brain, and the consequences in any one of these interactions could have a cascade-like effect that varied with the individual, the length of time on the drug, and so forth. But the one ironic consequence of these observations on the effect of giving people L-dopa was that the side effects themselves were soon regarded by psychiatrists as analogous to schizophrenia. The conclusion was drawn that the cause of schizophrenia was a disorder in dopamine metabolism—a sort of opposite to Parkinsonism. This issue is explored more fully in the next chapter.

The point isn't that L-dopa should not be used for the control of Parkinson's disease—it and variations of it remain some of the most effective treatments available. It is that introducing a drug into a complex system like the brain is a bit like throwing a wrench into the workings of any complex piece of machinery—there is no single consequence, but a lot of cogs get chewed up.

Even if the belief in magic bullets were better founded in biological reality, it is important to recognize that the social reality of the way in which drugs are used in general medical and psychiatric practice is far different from the neatly controlled studies of carefully matched patients that form the stuff of the clinical trials that give psychopharmacologists their scientific reputations and fill the pages of the scientific journals.

Lithium chloride was introduced, after some careful clinical trials, for the control of a relatively rarely diagnosed mental disorder, cyclical manic depression. Leaving aside the validity of the diagnosis of this condition, by the time lithium was available for general prescription, it was being prescribed in vast quantities not merely for the original disorder but now for depression, schizophrenia, and all stages in between. So widespread is its hospital use in Britain today that one psychopharmacologist has remarked that the concentrations of lithium in hospital effluent, recycled into the general drinking water, might soon reach a concentration high enough to produce a lithium toxicosis in the entire population of the country, as of course it is not removed by sewage treatment.

But medicalizing ideology will accept only those substances sanctioned by scientific orthodoxy and the drug companies. Lithium for depression, or dopamine antagonists for schizophrenia, or indeed drugs for "organic" diseases like multiple sclerosis are medically acceptable. On the other hand, when popular culture or uncertificated practitioners present their own magic-bullet solutions—vitamin C for colds, gluten-free diets for schizophrenia—or they suggest that the increasing incidence of inner-city depression is caused by lead in gasoline or paint, orthodoxy is scandalized; the experts' own techniques and theories have been turned against them. Popular magic-bullet remedies are no more—but no less—theoretically flawed than are those of the pharmaceutical industry. They are just as reductionist in inspiration. Perhaps we could regard them as refractions in popular culture of ruling ideologies, rather like working-class or black forms of Christianity. Like these religious ideologies, they are a contradictory mix of oppressive beliefs and critical opposition to ruling orthodoxies, whether of the priesthood or the pharmaceutical industry.

A consequence is that capitalism continuously tries either to disqualify or assimilate them. Vitamin C, for instance has been thoroughly assimilated in both the United States and Britain. In the United States the widespread use of expensive unleaded gasoline is a response to the critics which simply passes back to the consumer the cost of protecting his own health from corporate assault. The steady tendency toward the assertion of medical control over the use of popular psychotropic drugs runs as a strong thread through the history of medicine (for example, the medicalization of heroin and morphine over the last century).[54]

But if popular alternatives to medical orthodoxy threaten whole technologies, they can no longer simply be absorbed; they become a more critical challenge to capital and its experts. Those claiming that mental and bodily health can only come through radical dietary change threaten all of agribusiness. The argument that a major cause of cancer is environmental pollution from long-lasting and toxic chemicals generated by industry endangers much of the chemical industry. The claim that depression is the inevitable lot of women in a nuclear family threatens patriarchy.

Cures for the widespread social distress and individual existential despair of advanced patriarchal capitalist or so-called socialist societies cannot be found merely by manipulating the biology of the individual members of that society. Yet the nature of the society in which we live profoundly affects our biology as well as our behavior. In a healthier and more socially just society, even though pain, illness, and death will never be eliminated, our own individual biologies will nonetheless be different and healthier.

CHAPTER EIGHT

SCHIZOPHRENIA: THE CLASH OF DETERMINISMS

The Medicalization of Madness

The scale of diagnosed mental illness is now prodigious. In Britain, for example, some 170,000 patients are admitted each year to hospitals for various categories of "mental illness" (and another 16,000 for "mental handicap"). Mental illness patients these days are discharged quite soon, so there are only some 80,000 in hospitals at any one time. Mental handicap patients stay longer—there are almost 47,000 in hospitals at any one time. Put another way, one in twelve men and one in eight women in the United Kingdom—the proportions are similar in the United States—will now go to a hospital at some point in their lives to be treated for mental illness.[1] Yet the medical colonization of madness is a rather recent phenomenon; only in the last two centuries has madness been regarded as a medical matter at all.[2]

These figures are not static, reflecting as they do changing social definitions of wellness and illness, assumptions about the necessity for and most appropriate form of treatment, and so forth. Thus in recent years there have been dramatic changes in the mental hospital population. There has been an increase in the number of admissions to hospitals, but the average length of stay in hospitals has gone down. The result is a decline in the number of inpatients, that is, people confined to hospitals and regarded as unfit to leave for a period. Instead, more of those diagnosed as mentally ill are treated as outpatients outside the hospital ("within the community"—generally, that is, by their family) than previously. Perhaps the most striking example of such a change has been in Italy, which in 1978 passed a law closing *all* mental hospitals. From then on patients were to be treated in the community or as part of general hospital practice.

At an earlier time psychiatrists and neurologists chose to distinguish between "organic" and "functional" nervous disorders. In organic disorders there was something obviously and demonstrably wrong with the brain. There might be a lesion, or the aftermath of a stroke or toxic poisoning, or whatever. By contrast "functional" disorders—schizophrenia, the depressions, paranoia, and so on—were disorders of the mind, which could not be attributed to any obvious brain damage. We can see in this distinction a residue of the old Cartesian dualism, a split between body and mind functions. Some contemporary psychiatrists wish to maintain this position. In his numerous books polemicizing against contemporary institutional psychiatry, Thomas Szasz, for instance, argues that if schizophrenia were shown to have an associated biological dysfunction, it should be left to the compulsory medicalization of the state for treatment, but so long as it remained a disorder of the spirit with no clear biological component it should be the voluntary choice of the sufferer whether or not to visit his or her psychiatrist for paid therapy.[3]

But such a distinction is unacceptable to the dominant, full-blooded materialism of contemporary psychiatry. If there is a disordered mind, there must be associated with it some type of disordered molecular or cellular event in the brain. Further, the reductionist argument insists that there must be a direct causal chain running from the molecular events in particular brain regions to the most full-blown manifestations of existential despair suffered by the individual.

Present-day biological psychiatry divides disorders between neuroses, such as anxiety, and psychoses, of which schizophrenia is the outstanding example and the most common form of mental illness diagnosed today. The distinction offered between neurosis and psychosis is that in the former it seems as if sufferers perceive the same "real world" as do "normal individuals" but cannot react effectively and adaptively to it. By contrast, in the psychosis the individual's world ceases to be normal at all, at least for a considerable part of the time. Instead it is replaced by one in which large elements seem to be of the sufferer's own making, composed of fragments of the real world seen through a multifaceted distorting mirror. To the outside observer the psychotic is seen to be suffering from hallucinations and delusions.

But such definitions are inevitably uncertain. To begin with, they rest on a judgment about the meaning of normality. This involves comparing a given individual's behavior with that of his or her fellows in similar situations, or a person's behavior today with that on some previous occasion. It then becomes clear that definitions of normality are themselves time—and culture—bound. Joan of Arc—who heard voices which she claimed were those of angels telling her to crown the French Dauphin and drive out the English—became a heroine of the French nation. Later, long after her death, she was made a saint. Today she would almost certainly be diagnosed as schizophrenic, even though she might be spared burning at the stake. If an individual is cast into an apathetic despair about the likelihood of the world surviving nuclear holocaust through the eighties, or a woman in a northern English town is afraid to go out of her house at night for fear of being raped or murdered, how is one to judge that these are inappropriate responses compared with those of the less sensitive majority?

Clinicians often attempt to distinguish between "exogeneous" and "endogeneous" depressions. The former, it is claimed, are precipitated by events in the world outside the individual—a bereavement or loss of job for example—but sometimes even by events such as a promotion or moving to a new house. Endogenous depressions are said to be without obvious external precipitants and may recur cyclically at regular intervals, sometimes alternating with periods of exaggerated, frenzied cheerfulness (cyclical manic depression). In our present culture, depressions are often associated with important life cycle events (e.g. postpartum or postmenopausal depression). And even exogenous depressions may seem to take on a life of their own and fail to respond to the resolution of the initial precipitating cause. Women receive a higher proportion of diagnoses of depression and anxiety than men. The typical depressive consulting her family doctor is likely to be a middle-aged housewife.

Despite the textbook neatness of the distinctions between endogenous and exogenous depression and anxiety, the likelihood is that for most sufferers the distinctions are actually unclear; and in the clinical practice of most family doctors, rough-and-ready diagnostic criteria do not allow for much subtlety. In any event, once the diagnosis has been made the question is whether to try to normalize sufferers by persuading them that their despair or anxiety is uncalled for or by dulling it with drugs. It is immediately apparent that the relationship between the diagnosis of a behavior as illness and making judgments about appropriate and normal behavior is a very close one. It is at this point that the questions of cure and of control begin to intermingle, perhaps inextricably.

The Case of Schizophrenia

The diagnosis and treatment of schizophrenia are paradigms of the determinist mode of thinking, for this is the mental disorder on which more biochemical and genetic research has been lavished than any other, the one in which claims to have discovered *the* cause in a particular molecule or gene have been made most extensively. It is now so widely believed that psychiatry has proved the disorder to be biological that if the case fails here, where it is strongest, it must be even weaker elsewhere. But schizophrenia is interesting from another point of view as well, for in opposition to the biologizing tendencies of medical psychiatry there has grown up a strong countermovement in recent years. Antipsychiatry, in the hands of practitioners like R. D. Laing and theorists like Michel Foucault, has gone far in the opposite direction, almost to the point of denying the existence of a disorder or group of disorders diagnosable as schizophrenia at all. Thus in the case of schizophrenia we find precisely that clash of determinisms, on the one hand biological and on the other cultural, which we discussed in general in Chapters 3 and 4 and which it is one of the purposes of our book to transcend.

If the bulk of our effort here is directed toward the biochemical and particularly the genetic explanations offered for schizophrenia, this is because at the present time these explanations are so strongly entrenched in establishment psychiatry and medicine. We emphatically do not wish in this emphasis to be tipped over into an uncritical resuscitation of dualism, or cultural determinism like that of Laing or Foucault.

WHAT IS SCHIZOPHRENIA?

Schizophrenia literally means "split mind." The classic picture of a schizophrenic is of a person who feels in some fundamental way cut off from the rest of humanity. Unable to express emotion or interact normally or express themselves verbally in a way that is rational to most others, schizophrenics appear blank, apathetic, dull. They may complain that their thoughts are not their own or that they are being controlled by some outside force. According to the textbooks, dramatically ill schizophrenics appear not to be able to or wish to do anything for themselves—they take little interest in food, sexual activity, or exercise; they experience auditory hallucinations; and their speech seems rambling, incoherent, and disconnected to the casual listener. Some psychiatrists doubt whether schizophrenia is a single entity at all, or speak of core schizophrenia and a wider range of schizophrenia-like symptoms.

The idea of a single disease of schizophrenia may be a hangover from the nineteenth-century definition of madness—so-called dementia praecox—which preceded it. The diagnosis of schizophrenia in a patient with a given set of symptoms can vary between doctor and doctor and culture and culture. It is true that when matched and carefully controlled transnational surveys are done there is some concordance of diagnosis; however, in real life the diagnostic and prescribing practices of doctors and psychiatrists differ sharply from the more controlled procedures of clinical trials. Comparisons of figures in different countries have shown that the most frequent use of the diagnosis of schizophrenia occurs in the United States and the Soviet Union. Nonetheless, even in Britain, where it is defined in a somewhat narrower sense, up to 1 percent of the population is said to suffer from schizophrenia;[4] and 28,000—or 16 percent—of the admissions to hospital for mental illness in 1978 were for a diagnosis of schizophrenia or its related disorders.

Faced with the complex phenomena that result in a diagnosis of schizophrenia, the biological determinist has a simple question: What is it about the biology of the individual schizophrenic that predisposes him or her toward the disorder? If no obvious gross brain difference can be found, predisposition must lie in some subtle biochemical abnormality—perhaps affecting the connections between individual nerve cells. And the thrust of the determinist argument is that the causes for these abnormalities, although they might have been environmental, are most likely to lie in the genes.

The Drug Industry and Mental Illness

Hence the enthusiastic hunt, over many decades now, for the biochemically abnormal component in schizophrenia. How should this search be conducted? A standard pattern in the biologizing of human medicine has been to seek for experimental animals that show what appear to be analogous symptoms. Or the animals can be induced to manifest similar symptoms by damaging them in some way, infecting them, or treating them with drugs. In the case of mental disorders this approach is problematic. How could one recognize a schizophrenic cat or dog, even if the term had any meaning anyhow? Such difficulties have not entirely chilled the enthusiasm of the researchers. Experimental animals have been treated with drugs such as LSD and have been shown to become disoriented, to show abnormal fear reactions, or whatever. These may be interpreted as analogous to hallucination, and hence the effect of the drug is argued to be analogous to the assumed biochemical dysfunction in schizophrenia.

But such evidence is not very convincing, and most research is directed to a study of the biochemistry of the schizophrenic subjects themselves. Brain samples are rarely obtainable except postmortem, and so more readily accessible body materials—urine, blood, or cerebrospinal fluid—from certified schizophrenics are compared with those from control "normal" people with all the assiduity that the Roman augurs used to apply to the examination of animals' entrails. It is assumed that any biochemical abnormality in the brain will reflect itself in the production of abnormal metabolites in the blood, ultimately to be excreted in the urine.

When such approaches were first adopted several decades ago they soon began to show up large differences in the biochemistry of hospitalized schizophrenic patients from those of normals matched for sex, age, and so forth. But these differences turned out to be artifactual; nonschizophrenic hospitalized patients showed similar differences from the normal. The differences were eventually traced to the effects of long periods of eating poor hospital diets, or to the chemical-breakdown products of drugs that had been administered to the patients— or even to excessive coffee-drinking by hospitalized patients.

Even when proper care is taken to circumvent this problem by ensuring that the subjects studied have been kept off drugs for a period, that they have the same diet as their matched controls, and so forth, there remains a general methodological problem that cannot be avoided. Even if an abnormal chemical is found in the body fluids of a diagnosed schizophrenic compared with the best-matched of controls, one cannot infer that the observed substance is the cause of schizophrenia; it might instead be a consequence. The causal argument assumes that the substance is present, and, as a result, the disorder begins. A consequential argument says that first the disorder occurs and then as a result the substance accumulates. If an individual suffers an infection from a flu virus there is a considerable increase in the antibodies present in the blood and mucus of the nose—they are the body's defense mechanisms against the virus. The antibodies and the mucus haven't caused the infection, and one cannot readily deduce the actual causes simply by observing such consequences.

Such problems have made yet another approach more attractive to reductionist thinking: to observe the effects of pharmacological agents —drugs—on human behavior. If a drug induces schizophrenia-like behavior—for example, auditory hallucinations—then attempts will be made to conclude that the drug interferes with a biochemical process in the normal person which is damaged in the schizophrenic. Hence, for example, there was a period in the 1960s in which attempts were made to find links between LSD and schizophrenia on the grounds that users of LSD experienced hallucinations that might be seen as analogous to those of the schizophrenic. This logic, which argues backwards from the effect of a drug to the cause of a disease (*ex juvantibus* logic),[5] is plainly a risky procedure, both for the logician and for the patient. As we have emphasized in the case of L-dopa, no drug has a single site of action. Foreign chemicals introduced into the body are not magic bullets.[6]

Yet such thinking has dominated more than thirty years of research on the biochemistry of schizophrenia, generated endless research papers, made scientific and medical reputations, and brought incidental substantial profit to the big drug firms. The history of thinking among biochemists about schizophrenia over the period is inextricably intertwined with that of the pharmaceutical industry, for which psychotropic drugs have been one of the biggest money spinners. One in five of drugs issued in the British National Health Service in 1979 was for a drug acting on the central nervous system. Hoffmann-La Roche earns nearly $1 billion a year worldwide from its sales of Valium. It is estimated that chlorpromazine, introduced in 1952 for the control of long-stay hospitalized schizophrenics and related patients, had been administered to 50 million people worldwide within the first ten years of its use.

There is still another twist to the spiral of interdependence of the drug industry and the diagnosis of mental illness. With prolonged use of drugs, a whole new range of disorders has become apparent. Substances intended to cure one problem generate another, and the growth in such iatrogenic (medically induced) disorders is serious and disturbing. This is particularly the case for the major tranquilizers like chlorpromazine. There has been a slow recognition in the last decade or so of the disorder category known as tardive dyskinesia, apparent particularly among hospitalized patients who have been long users of chlorpromazine. The symptoms, which include characteristic motor disabilities and uncontrollable gestures (for instance, movements of the mouth), do not necessarily disappear when the patient is taken off the drug. There are reports that between 10 and 40 percent of those who regularly use major tranquilizers may suffer from tardive dyskinesia, and about 50 percent of those who get the disorder will have some irreversible consequential brain damage. Nor are there at present any drugs to combat these effects, though tardive dyskinesia has become a prolific spin-off area for neurobiological research.[7]

It would be wearisome and unnecessary to recount in detail the history of research into the biochemistry of schizophrenia over the past thirty years. Almost every biochemical substance known to be present in the brain has, within two or three years of its introduction into the biochemical dictionary, been studied for possible involvement in schizophrenia by clinical scientists with the hope of a breakthrough in their hearts and with grant money (often from drug companies) burning holes in their pockets.

We do not in any way wish to minimize the enormous difficulties faced in clinical research. The desire for a solution to the problem of schizophrenia is real and great, and the insistence on a biological mode of explanation that will enable effective drugs to be developed is part of a pressurizing culture to which clinical research is responding. Drugs that alleviate symptoms, like the use of aspirin for toothache, may be worth developing even if they tell nothing about the causes of the disorder. The multiplicity of drugs (and formulations of drugs) is an aspect of the way the pharmaceutical companies work in a field where knowledge of patent law is as important as clinical skills. The problem is that of confounding the effect of a drug with the offer of an explanation, the alleviation of suffering with a cure for the disease.

Among the claims for causative factors in schizophrenia made since the 1950s we may point to: abnormal substances secreted in the sweat of schizophrenics; injection of the blood serum of schizophrenics into other, normal subjects inducing abnormal behavior; and the presence of abnormal enzymes in red blood cells and blood proteins. Between 1955 and the present day, conflicting research reports have claimed that schizophrenia is caused by disorders in serotonin metabolism (1955); noradrenaline metabolism (1971); dopamine metabolism (1972); acetylcholine metabolism (1973); endorphin metabolism (1976); and prostaglandin metabolism (1977). Some molecules, such as the amino acids glutamate and gamma-amino-butyric acid, came into fashion in the late 1950s, fell into neglect, and now, in the 1980s, have come back into fashion once more.[8]

Most of the substances referred to above are brain chemicals known to play a part in the transmission of nerve impulses between cells. This points to the main idea running through all such research. The notion is that in some way, in schizophrenia, messages between cells in those regions of the brain concerned with information processing and with affect become scrambled, resulting in inappropriate responses. The evidence for any and all of the various molecular disorders is based on a combination of the types of methodologies and logic described earlier. Rarely have results obtained by one group of researchers been confirmed by another group of researchers in a different group of patients. Rarely has any resolution of conflicting claims been attempted. Rarely has any concern been expressed by the enthusiastic clinical researchers that schizophrenia might be associated with many different biochemical effects, or indeed that many different types of biochemical change might lead to or be generated by the same behavioral outcomes.

The Genetics of Schizophrenia

The statement that the brain of a person manifesting schizophrenia shows biochemical changes compared with that of a normal person may be no more than a reaffirmation of a proper materialism that insists on the unity of mind and brain. But the ideology of biological determinism goes much deeper than this. It is, as we have reiterated, linked to an insistence that biological events are ontologically prior to and cause the behavioral or existential events, and hence to a claim that if brain biochemistry is altered in schizophrenia, then underlying this altered biochemistry must be some type of genetic predisposition to the disorder. By 1981 psychologists were claiming to be able to detect potential schizophrenics when they are only three years old—up to fifty years before the disease manifests itself. The claim, made by Venables to a meeting of the British Association for the Advancement of Science, is based on a survey of three-year-olds in Mauritius; "potentially abnormal" children were said to show "abnormal autonomic responses."[9]

Push the diagnosis back beyond the three-year-old and we are soon with embryo or gene. But the hunt for a genetic basis for schizophrenia goes far beyond an interest in therapy, as there is no way in which the mere demonstration of a genetic basis for the disorder would aid in its treatment.* As we have seen, the lineage of the effort to find genetic predispositions runs back through the eugenic thinking of the 1930s and 1920s, with its belief in genes for criminal degeneracy, sexual profligacy, alcoholism, and every other type of activity disapproved of by bourgeois society. It is deeply embedded in today's determinist ideology. Only thus can we account for the extraordinary repetitive perseverance and uncritical nature of research into the genetics of schizophrenia. Whatever such research may say about the disorder it proposes to explain, an examination of the claims of its protagonists says a very great deal about the intellectual history of our contemporary determinist society, and hence is worth analyzing in some detail.

The belief that schizophrenia has a clear and important genetic basis is now very widely held. The father of psychiatric genetics, Ernst Rüdin, was so convinced of this that, arguing on the basis of statistics collected by his co-workers, he advocated the eugenic sterilization of schizophrenics. When Hitler came to power in 1933, Rüdin's advocacy was no longer merely academic. Professor Rüdin served on a panel, with Heinrich Himmler as head, of the Task Force of Heredity Experts who drew up the German sterilization law of 1933.

Perhaps the most influential psychiatric geneticist in the English-speaking world was a student of Rüdin's, the late Franz Kallmann. The blizzard of statistics published by Kallmann seemed to indicate conclusively that schizophrenia was a genetic phenomenon. From his study of a thousand pairs of affected twins, Kallmann concluded that if one member of a pair of identical twins was schizophrenic there was an 86.2 percent chance that the other would be also. Further, if two schizo-

---

*These words were true when we wrote them. However, reductionist science moves faster than the Gutenberg technology of book production. For if it were the case that there were schizophrenia-producing genes, then techniques that excised those abnormal genes from the genome of affected individuals and replaced them with their normal alleles would presumably prevent the expression of the disorder. If schizophrenia were a single or even a two- or three-gene defect, such techniques are not wholly beyond the reach of contemporary molecular genetics—what is sometimes called genetic engineering. There are serious research programs now under way in several laboratories to make gene libraries from schizophrenics and isolate and clone the "schizophrenic genes" with a view to studying their possible replacement. Granted the reductionist premise, the therapeutic logic would be impeccable. And if one can have schizophrenic urine, why not, indeed, schizophrenic genes?

phrenic parents produced a child, there was a 68.1 percent chance that the child would be schizophrenic. These figures led Kallmann to argue that schizophrenia could be attributed to a single recessive gene.

The particular genetic theory espoused by Kallmann has made it possible for latter-day psychiatric geneticists to attempt a spectacular rewriting of their history. Thus, in a recent textbook the following note appears: "Kallmann's [theory] was apparently not based solely on his data. His widow has indicated that Kallmann advocated a recessive model because he could then argue convincingly against the use of sterilization to eliminate the gene. As a Jewish refugee, Kallmann was very sensitive to this issue and afraid of the possible social consequences of his own research."[10] The point here is that if a disease such as schizophrenia is caused by a recessive gene, many carriers of the gene will not themselves display symptoms. Thus, sterilization merely of those who do show symptoms would be inefficient and would fail to eliminate the disease.

The picture of Kallmann as a bleeding-heart protector of schizophrenics, adjusting his scientific theories to mirror his compassion, is grotesquely false. The first Kallmann publication on schizophrenia is in a German volume edited by Harmsen and Lohse that contains the proceedings of the frankly Nazi International Congress for Population Science.[11] There, in Berlin, Kallmann argued vigorously for the sterilization of the apparently healthy relatives of schizophrenics, as well as of schizophrenics themselves. This was necessary, according to Kallmann, precisely because his data indicated that schizophrenia was a genetically recessive disease. Two Nazi geneticists, Lenz and Reichel, rose to argue that there were simply too many apparently healthy relatives of schizophrenics to make their sterilization feasible.

The eugenicist views of Kallmann were not confined to obscure Nazi publications but were also made widely available in English after his arrival in the United States in 1936. In 1938 he wrote of schizophrenics as a "source of maladjusted crooks, asocial eccentrics, and the lowest type of criminal offenders. Even the faithful believer in . . . liberty would be much happier without those. . . . I am reluctant to admit the necessity of different eugenic programs for democratic and fascistic communities . . . there are neither biological nor sociological differences between a democratic and a totalitarian schizophrenic."[12]

The extremity of Kallmann's totalitarian passion for eugenic sterilization was clearly indicated in his major 1938 text. Precisely because of the recessivity of the illness, it was above all necessary to prevent the reproduction of the apparently healthy children and siblings of schizophrenics. Further, the apparently healthy marriage partner of a schizophrenic "should be prevented from remarrying" if any child of the earlier marriage is even a suspected schizophrenic, and even if the second marriage is with a normal individual.[13]

These views of the future president of the American Society for Human Genetics are so bloodcurdling that one can sympathize with the efforts of present-day geneticists to misrepresent or to suppress them. They have not, however, suppressed the mountains of published statistics with which Kallmann attempted to prove that schizophrenia (like tuberculosis and homosexuality) was a hereditary form of degeneracy. Those figures are presented to students in today's textbooks as the fruits of impartial science. We begin our review of the data concerning the genetics of schizophrenia with a detailed examination of Kallmann's work, which should make clear that Kallmann's figures cannot be regarded seriously.

Kallmann's Data

The Kallmann data were collected under two very different sets of circumstances. The earlier data, published in 1938, were based upon the records of a large Berlin mental hospital. Working with records from the period 1893–1902, Kallmann made an "unambiguous diagnosis" of schizophrenia in 1,087 index cases. To make these diagnoses it was necessary to ignore "earlier diagnoses or the contemporary notes on hereditary taint conditions in the family of the patient." Then Kallmann attempted to locate, or to acquire information about, relatives of the index cases—many of whom were long since dead. That task often involved

formidable difficulties . . . we were dealing with inferior people. . . . They sometimes escaped our search for years. . . . Quite a few were bad-humored . . . we had to overcome the suspicion with which certain classes regarded any kind of official activity. . . . Whenever we encountered serious opposition we found ourselves to be dealing with either officials and members of the academic world, or people with exaggerated suspicions, schizoid types, and possible schizophrenics . . . our private sources of information were amplified from the records of police bureaus. . . . In making inquiries about people already dead or living too far away, we employed . . . local bureaus and trusted agents.[14]

With information gathered in this way, Kallmann felt able to diagnose the relatives of the index cases, and thus to report the probability of schizophrenia for each type of relative. The rates reported by Kallmann in this German sample are reproduced in the left-hand column of Table 8.1. The reported rates, it should be noted, were "age-corrected." That was necessary because some of the relatives were quite young and might develop schizophrenia as they grew older. The arbitrary correction employed by Kallmann can sometimes produce rates in excess of 100 percent.

The second set of data collected by Kallmann came from a very different sample, studied in New York State. The index cases were now individuals who were schizophrenic twins who had been admitted to public mental hospitals. When Kallmann reported in 1946, there were 794 such index cases.[15] By 1953, the number had increased to 953. There were, of course, some identical (MZ) twins, and some fraternal (DZ) twins. Thus, by obtaining information about the co-twins of index cases, Kallmann could report the probability that both members of a pair were schizophrenic. That probability is called the "pairwise concordance rate." The age-corrected concordances were reported for different types of twins, along with the corrected morbidity rates for various types of relatives. These had been determined by collecting information about the relatives of the twin index cases. There was virtually no information given about the procedures employed in this massive study, but Kallmann wrote that "classification of both schizophrenia and zygosity were made on the basis of personal investigation and extended observation." This obviously allowed for "contaminated diagnosis." That is, the decision as to whether or not a co-twin was said to be schizophrenic could be influenced by the decision as to whether the twin pair was MZ or DZ and vice versa. The Kallmann 1946 data, and the even more sketchily reported data of 1953,[16] are also presented in Table 8.1.

TABLE 8.1 / Age-corrected morbidity rates for schizophrenia, as reported by Kallmann

Relationship to Index Case:	Berlin, 1938	New York, 1946	New York, 1953
MZ Twin	—	85.8	86.2
DZ Twin	—	14.7	14.5
Parents	10.4	9.2	9.3
Children	16.4	—	—
Full siblings	11.5	14.3	14.2
Half-siblings	7.6	7.0	7.1
Grandchildren	4.3	—	—
Nephews, nieces	3.9	—	—
Step-siblings	—	1.8	1.8
Spouse	—	2.1	—

These data are obviously consistent with an overwhelming genetic determination of schizophrenia—particularly the remarkable rate of 86 percent among MZ twins. Where direct comparisons can be made, the change of countries and of eras—as well as the switch to relatives of twin index cases—has had little effect on the reported figures.

The correspondence between Kallmann's theoretical expectations and the results he discovered is sometimes quite remarkable. Thus, in 1938 Kallmann indicated that the work of earlier twin researchers suggested that schizophrenia manifested itself, even among those with the full genetic predisposition, only about 70 percent of the time.[17] That meant, according to Kallmann's single recessive gene theory, that 70 percent of the children of two schizophrenic parents should themselves be schizophrenic. The Kallmann data indicated that the expectation of schizophrenia in the offspring of two schizophrenics was precisely 68.1 percent. That result, of course, nicely validated Kallmann's theory. Four other studies of the children of two schizophrenic parents suggest a risk of only between 34 and 44 percent.[18]

Kallmann stressed repeatedly that, in his data, the "morbidity figure for the siblings . . . corresponds perfectly with the concordance rate for two-egg twin pairs, whose chance of inheriting a similar genotypical combination is exactly the same as that for any ordinary pair of brothers and sisters".[19] The same close correspondence was described as a notable finding in 1953. We shall soon see, however, that—as an embarrassment to a simple genetic theory—other investigators have not found the close correspondences of data with theory routinely detected by Kallmann.

There are many similarities between the roles of Franz Kallmann in schizophrenia research and of Cyril Burt in IQ research. The two men each believed passionately in the genetic determination of human behavior. While Kallmann fulminated against the dysgenic threat posed by schizophrenics, Burt—also a eugenicist—was deeply concerned by the threat of dysgenic reproduction by low-IQ people. The two men each gathered by far the most massive sets of data collected in their fields. The two men each failed to describe with any adequacy at all their methods and procedures. The results reported by each were incredibly consistent with simple genetic theories—far more so than the data collected by other investigators. That happy coincidence enabled Kallmann to argue for "eugenic-prophylactic measures" against the families of the mentally ill, as it enabled Burt to argue against wasting educational resources on those with low IQ scores. As we showed in Chapter 5, there is now universal agreement that Burt's data were fraudulent and must be discarded. The same, however, is not true of Kallman's data. In fact, they have been defended vigorously against untoward insinuations. As Shields and his colleagues put it, these could only be made because "the abbreviated manner in which Kallmann reported his results left him more open than he would otherwise have been to criticism."[20]

The research conducted by others who followed Kallmann has in any event made it clear that his extraordinarily high figures cannot be repeated. The Kallmann data are still presented, unblushingly, in purportedly serious reviews of research, but they are now counterbalanced by more recent and more modest results. Perhaps the chief harm brought about by Kallmann's deluge of incredible and poorly documented data was to create a climate in which the findings of subsequent workers seemed so reasonable and moderate that they escaped serious critical scrutiny. Thus, Kallmann's data have faded from the body of acceptable evidence, but the belief for which he was largely responsible

—that a genetic basis for schizophrenia has been clearly established—still remains powerful in and out of science.

FAMILY STUDIES

There are basically three kinds of inquiries that attempt to demonstrate a genetic basis for schizophrenia: family studies, twin studies, and adoption studies. There is no need to spend much time on the first. The simple idea behind them is that if schizophrenia is inherited, the relatives of schizophrenics are likely to display the disease as well. Further, the more closely related a person is to a schizophrenic, the more likely it should be that the person will be affected. The problem is, of course, that these predictions would also follow from a theory that maintained that schizophrenia was environmentally produced. There is an obvious tendency for close relatives to share similar environments.

For what such data are worth, the major compilation of family studies seems to have been made by Zerbin-Rüdin.[21] The compilation was presented to English-readers in "simplified form" by Slater and Cowie.[22] Their table indicates, e.g., that fourteen separate studies yield a 4.38 percent expectation of schizophrenia among the parents of schizophrenic index cases. The expectation among sibs, in ten studies, was 8.24 percent; and among children, 12.31 percent in five studies. For uncles and aunts, grandchildren, and cousins the figures were all under 3 percent, but still higher than the expected 1 percent.

The exactness of these figures, however, is more apparent than real. The same basic set of studies was also summarized by Rosenthal in 1970.[23] The relatives diagnosed in these studies, Rosenthal noted, had often been dead for many years. The studies are quite old, and methods of diagnosis and of sampling are not always spelled out. The combined figures are dominated by Kallmann's massive samples and by data gathered by other members of Rüdin's "Munich school." The Rosenthal tables make clear a fact that is obscured by the Slater and Cowie summary. There are vast differences in the rates of schizophrenia reported in different studies. For parents of index cases, reported risks range from 0.2 percent (lower than in the population at large) to 12.0 percent. For sibs, the range is between 3.3 and 14.3 percent. The risk for sibs is in one study twenty-nine times larger than that for parents; but in another the risk for parents is 1½ times larger than that for sibs. These studies at best demonstrate what nobody would have contested. There is at least a rough tendency for diagnosed schizophrenia to "run in families."*

TWIN STUDIES

As described in Chapter 5, the basic logic of twin studies depends upon the fact that while MZ twins are genetically identical, DZ twins on average share (like ordinary siblings) only half their genes. Thus, if a trait is genetically determined, one would obviously expect MZs to be concordant for that trait more often than DZs. The major logical problem with twin studies is that MZ twins, who typically resemble one another strikingly in appearance, are treated much more similarly than are DZs by parents and peers. There is abundant evidence (discussed in Chapter 5) that the environments of MZs are very much more similar than those of DZs. (Twin studies typically compare concordance rates among MZs, who are always of the same sex, with concordance rates among same-sexed DZs.) The demonstration that concordance is higher among MZs does not necessarily establish a genetic basis for the trait in question. Perhaps the difference is due to the greater environmental similarity of MZs. We shall soon discuss evidence which indicates that this possibility is not at all farfetched.

Well-designed twin studies should take as their index cases all schizophrenic twins admitted to a particular hospital during a particular time period. The alternative—feasible in small Scandinavian countries, which maintain population registers—is to start with the entire population of twins and to locate index schizophrenic cases. With either technique, a number of procedural problems are inevitable. The co-twins of index cases are often dead or unavailable for personal examination. Thus, informed guesses often must be made both about whether a given pair is MZ or DZ, and whether or not the co-twin is schizophrenic. The guesses are typically made by the same person, opening the way for contaminated diagnoses. There is sometimes an

---

*Even this modest conclusion is not unchallenged in the literature. Two studies in the United States found rates of schizophrenia among the first-degree relatives of schizophrenics which were scarcely above the rate in the general population.[24]

effort to have blind diagnoses made of individual cases by independent judges, working from written case histories.[25]

The case histories, however, contain selective material gathered and prepared by investigators who were not themselves "blind." Further, the case records of those twins who have in fact been hospitalized—and their diagnoses—had been written up by doctors who questioned the ill twins in detail about possible taint in their family lines. The diagnosis of schizophrenia, as should by now be clear, is by no means a cut-and-dried affair. The fact that a person's relative may have suffered from schizophrenia is often used to help doctors make a diagnosis.

The biases that contaminate twin studies stand out clearly from an attentive reading of the published case history materials. The very first case described by Slater in 1953 is the story of Eileen, a hospitalized schizophrenic, and of her identical twin, Fanny. Eileen had been hospitalized in 1899, "suffering from acute mania," and died in the hospital in 1946. With Eileen as the index case, Slater's task was to investigate the mental status of Fanny, who died, aged seventy-one, in 1938. We are told by Slater:

While still in the twenties she had a mental illness, of which no details are available. . . . Fanny in [1936] proved very difficult to examine . . . so that only the barest details were obtainable. She suppressed all mention of her own mental illness in early years, which fact was obtained from the history of her twin sister given at the time of her admission to hospital. Though there was no sign of any present schizophrenic symptoms, this suspicion and reserve are such as are commonly found as sequelae of a schizophrenic psychosis. Unfortunately, no facts are obtainable about the nature of her past mental illness, but the probabilities are very greatly in favour of it having been a schizophrenic one . . . she made a fairly complete and permanent recovery . . . though psychologically her reserve and lack of frankness suggest that the schizophrenia was not entirely without permanent after-effect. . . . According to her daughter-in-law, who had not heard of her mental illness, she led a hard life. Neither her family nor the neighbours noticed anything odd about her.[26]

These MZ twins, according to Slater, were concordant for schizophrenia. The only evidence that Fanny had once suffered from schizophrenia was her twin's assertion—while "suffering from acute mania" in 1899—that Fanny had had some kind of mental illness. Fanny herself, in 1936, was difficult and suppressed all mention of her illness. That lack of frankness, Slater noted, was typical of recovered schizophrenics, who otherwise appear normal. Fanny's dead identical twin had clearly been schizophrenic. For Slater this made it obvious that Fanny's supposed mental illness fifty years earlier had been schizophrenia. Fanny's neighbors and family, unlike Slater and other students of the Munich school, had not the wit to detect Fanny's schizophrenia.

Consider now the first pair of discordant DZ twins described by Gottesman and Shields in their 1972 study. Twin A was a hospitalized schizophrenic. What about Twin B? "No psychiatric history. Family unwilling for him to be contacted for Twin Investigation. . . . The pair differs from most in that neither twin was seen by us." The investigators concluded that Twin B was normal; and six blind judges, pondering a case study summary prepared by the investigators, unanimously agreed that Twin B was free of psychopathology. With DZ Pair 16 of the same study, all judges again agreed that the co-twin was normal, making the pair discordant. The diagnosis of the co-twin had not been made under ideal conditions: "He refused to be seen for the Twin Investigation, remaining upstairs out of sight, but his wife was seen at the door. . . . He was regarded as a healthy, levelheaded, solid happy person." That might in fact be the case—but few will agree that diagnoses of co-twins made in this way are solid or levelheaded.

Problems of this sort affect all twin studies, and that should be borne in mind as we review the results reported by various investigators. To obtain reasonable estimates of concordance rates, it seems sensible to require that a study contain at least twenty pairs of MZ and twenty pairs of same-sexed DZ twins. There have been seven such studies, and their results are summarized in Table 8.2.

The table presents raw, pairwise concordance rates, without any age correction. Two sets of rates are given for each study, one narrow and one broad. The narrow rates are based on the investigator's attempt to apply a relatively strict set of criteria when diagnosing schizophrenia. The broad rates include as concordant cases in which one twin is described as "borderline schizophrenic" or as "schizo-affective psychosis" or a "paranoid with schizophrenic-like features." The tabled concordance rates, it should be noted, depend upon the different investigators' varying sets of diagnostic criteria. They have not been concocted ad hoc by us.

TABLE 8.2. / Reported concordance rates

Study	"NARROW" CONCORDANCE: % MZs	"NARROW" CONCORDANCE: % DZs	"BROAD" CONCORDANCE: % MZs	"BROAD" CONCORDANCE: % DZs
Rosanoff et al., 1934[27] (41 MZs, 53 DZs)	44	9	61	13
Kallmann, 1946[15] (174 MZs, 296 DZs)	59	11	69	11–14
*Slater, 1953[26] (37 MZs, 58 DZs)	65	14	65	14
Gottesman and Shields, 1966[28] (24 MZs, 33 DZs)	42	15	54	18
Kringlen, 1968[29] (55 MZs, 90 DZs)	25	7	38	10
Allen et al., 1972[30] (95 MZs, 125 DZs)	14	4	27	5
Fischer, 1973[31] (21 MZs, 41 DZs)	24	10	48	20

*There is no simple way to derive separate narrow and broad concordance rates for Slater.

The table makes clear that in all studies concordance is higher for MZ than for DZ twins. But it is also clear that the concordance reported for MZs is much higher in the three older studies than in the four more recent ones. There is in fact no overlap between the two sets of studies. For narrow concordance, the average has plunged from 56 to 26 percent for MZs; for DZs, the corresponding averages are 11 and 9 percent. For broad concordance, MZ rates have dropped from 65 to 42 percent, while the DZ rate remained at a constant 13 percent. These average values, which weight all studies equally, should not be taken too literally. The data do make clear, however, that even in genetically identical MZs environmental factors must be of enormous importance. The concordance for MZs reported by modern researchers, even under the broadest criteria, does not remotely approach the preposterous 86 percent figure claimed by Kallmann.

Those who perform such studies still claim, however, that the higher concordance observed among MZs—a unanimous finding—demonstrates at least some genetic basis for schizophrenia. We have already noted that MZs not only are genetically more similar than DZs but also experience much more similar environments than do DZs. The environmental similarity, no less than the genetic similarity, might plausibly account for the higher concordance of MZs.

There are in fact some simple and critical tests that can be made of this environmental hypothesis. There is no doubt that DZ twins experience more similar environments than do ordinary siblings. The DZ twins, however, are genetically no more alike than are ordinary siblings—they are only siblings who happen to have been born at the same time. Thus, from an environmental viewpoint—and only from such a viewpoint—we would expect concordance among DZs to be higher than among ordinary sibs. There have been a number of studies that reported rates of schizophrenia concordance among DZ twins, as well as rates among siblings of the twins. The results of all such studies are summarized in Table 8.3.

Though the reported differences are very small in the early studies, all studies agree in showing a higher concordance rate among DZs than among sibs. Within more modern studies, the difference is often statistically significant, with the risk for DZs reported as two or three times that for sibs. When we note that similarity of environment can double or triple the concordance of DZs above that of sibs, it seems entirely plausible to attribute the still higher concordance of MZs to their still greater environmental similarity.

The same kind of point can be demonstrated by comparing the concordance rates of same-sexed and of opposite-sexed DZs. Though both types of DZ twins are equally similar genetically, it is obvious that same-sexed pairs experience more similar environments than do opposite-sexed pairs. The available data, summarized in Table 8.4, again support the environmentalist expectation. There have been statistically significant differences reported by several investigators, always indicating a higher concordance among same-sexed twins. The results of the one study that appears to reverse the otherwise universal trend were not statistically significant.

Consider, finally, some implications of a finding casually reported by Hoffer and Pollin.[35] Those authors studied the hospital records of the American war veteran twins later reported on by Allen et al. Several hundred diagnosed schizophrenic twins were located by searching through records, but the twins were not personally examined by the

TABLE 8.3 / Reported risks for DZ twins and sibs

	% DZs	% Sibs
Luxenburger, 1935[32]	14.0	12.0
Kallmann, 1946[15]	14.7	14.3
*Slater, 1953[26]	14.4	5.4
Gottesman and Shields, 1972[25]	9.1	4.7
*Fischer, 1973[31]	26.7	10.1
Kringlen, 1976[29]	8.5	3.0

*Probability that the differences between DZs and sibs are due only to sampling error is less than 0.01 %.

investigators. Thus, to determine whether a twin pair was MZ or DZ, questionnaires were mailed to all twins, asking whether they looked as much alike as two peas in a pod, whether they were confused for each other, etc. There were many occasions when only one twin of a discordant pair returned the questionnaire. When the twin returning the questionnaire had been diagnosed as schizophrenic, 31.3 percent gave answers indicating that they were MZ. When the answering twin was not the diagnosed schizophrenic, only 17.2 percent indicated that they were MZ. The difference is statistically significant, and it was produced by an unrealistically small proportion of MZs among the nonschizophrenic twins.

That is easily understandable. When you are normal and your twin is schizophrenic, you are well advised to tell twin investigators and other authorities that you are not a carbon copy of your twin—even if you really are MZs. To admit that you are the MZ twin of a schizophrenic is clearly to invite a similar diagnosis—even, perhaps, sterilization—for yourself. We recall that in all the twin studies some decisions about zygosity are made on the basis of questions put to nonaffected twins and to their relatives. With a little sensitivity to the real lives of people, we must recognize an all-too-human tendency to deny that the nonaffected MZ twins of schizophrenics really are identical. This must be still another source of error, tending to remove some discordant pairs from the MZ and into the DZ category. That, of course, artificially inflates the difference in concordance rates between

TABLE 8.4 / Concordance in same- and opposite-sexed DZ twins

	% Same-Sexed	% Opposite-Sexed
*Rosanoff et al., 1934 (53 SS, 48 OS)[27]	9.4	0.0
Luxenburger, 1935[32]	19.6	7.6†
*Kallmann, 1946 (296 SS, 221 OS)[15]	11.5	5.9
*Slater, 1953 (61 SS, 54 OS)[26]	18.0	3.7
Inouye, 1961 (11 SS, 6 OS)[33]	18.1	0.0
Harvald and Hauge, 1965 (31 SS, 28 OS)[34]	6.5	3.6
Kringlen, 1968 (90 SS, 82 OS)[29]	6.7	9.8

*Probability that differences between same- and opposite-sexed twins are due only to sampling error is less than 0.05 %.

†Estimated.

MZs and DZs. There is little wonder in the fact that even psychiatric geneticists have not found twin studies to be wholly convincing, and have turned to studies of adoption. The adoption studies, in theory at least, might be able to disentangle genetic from environmental effects in a way that twin studies cannot.

ADOPTION STUDIES

The basic procedure of adoption studies is to begin with a set of schizophrenic index cases, and then to study the biological relatives from whom they have been separated by the process of adoption. Thus —at least in theory—the index case and his or her biological relatives have only genes, and not environment, in common. The question of interest is whether the biological relatives of the index cases, despite the lack of shared environments, display an increased incidence of schizophrenia. To answer that question it is necessary to compare the rate of schizophrenia among the biological relatives with the rate observed in some appropriate control group.

The adoption studies carried out in Denmark in recent years by a collaborative team of American and Danish investigators have had enormous impact. To some critics who could detect the methodological weaknesses of twin studies, the Danish adoption studies appeared to establish the genetic basis of schizophrenia beyond any doubt. The eminent neuroscientist Solomon Snyder referred to these studies as a landmark "in the history of biological psychiatry. It's the best work that's been done. They take out all the artifacts in the nature vs. nurture argument."[36] Paul Wender, one of the authors of the studies, was able to announce: "We failed to discover any environmental component. . . . That's a very strong statement."[37] Though Wender's total excision of environmental factors is extreme, the Danish studies have been universally accepted as an unequivocal demonstration of an important genetic basis for schizophrenia. Clearly these studies require detailed critical examination.

Though they have been described in many separate publications, there are basically two major Danish adoption studies. The first, with Kety as senior investigator, starts with adoptees as the schizophrenic index cases and examines their relatives. The second, with Rosenthal as senior investigator, starts with schizophrenic parents as index cases and examines the children whom they gave up for adoption.

The study that began with adoptees as index cases was first reported by Kety in 1968.[38] Based on Copenhagen records, the investigators located thirty-four adoptees who had been admitted to psychiatric hospitals as adults and who could be diagnosed from the records as schizophrenics. For each schizophrenic adoptee a control adoptee who had never received psychiatric care was selected. The control was matched to the index case for sex, age, age at transfer to the adoptive parents, and socioeconomic status (SES) of the adoptive family.

The next step was to search the records of psychiatric treatment for all Denmark, looking for relatives of both the index and control cases. Those who searched the records did not know which were the relatives of index cases and which were the relatives of controls. Whenever a psychiatric record was found, it was summarized and then diagnosed blindly by a team of researchers who came to a consensus. The relatives were not at this stage personally examined.

The researchers traced 150 biological relatives (parents, sibs, or half-sibs) of the index cases, and 156 biological relatives of the controls. The first point to note is one not stressed by the authors: There were virtually no clear cases of schizophrenia among the relatives either of the index or of the control cases. To be precise, there was one chronic schizophrenic among the index relatives and one among the controls. To obtain apparently significant results the authors had to pool together a "schizophrenic spectrum of disorders." The spectrum concept lumps into a single category such diagnoses as chronic schizophrenia, "borderline state," "inadequate personality," "uncertain schizophrenia," and "uncertain borderline state." With such a broad concept, 8.7 percent of the biological relatives of index cases and 1.9 percent of the biological relatives of controls were diagnosed as displaying spectrum disorders. There were nine biological families of index cases in which at least one spectrum diagnosis had been made, compared to only two such families among the controls. That difference is the supposed evidence for the genetic basis of schizophrenia. Without the inclusion of such vague diagnoses as "inadequate personality" and "uncertain borderline schizophrenia" there would be no significant results in the Kety study.

From the Kety data of 1968 it is possible to demonstrate that such vague diagnoses—falling within the "soft spectrum"—are not in fact associated with schizophrenia. Among the sixty-six biological families reported on in 1968 there were a total of six in which at least one "soft" diagnosis had been made.* There was *no* tendency for such diagnoses to occur any more frequently in families in which definite schizophrenia had been diagnosed than in other families. However, the "soft spectrum" diagnoses very definitely tended to occur in the same families in which "outside the spectrum" psychiatric diagnoses had been made—that is, such clearly nonschizophrenic diagnoses as alcoholism, psychopathy, syphilitic psychosis, etc. There were "outside the spectrum" diagnoses in 83 percent of the families containing "soft spectrum" diagnoses, and in only 30 percent of the remaining families— a statistically significant difference. Thus it appears that the Kety et al. results depend upon their labeling as schizophrenia vaguely defined behaviors that tend to run in the same families as do alcoholism and criminality—but which do not tend to run in the same families as does genuine schizophrenia. However, it remains the case that these frowned-upon behaviors did occur more frequently among the biolog-

*We here include as "soft" diagnoses the two least certain diagnoses employed by Kety et al.—their D-3 diagnosis ("uncertain borderline") and their C diagnosis ("inadequate personality").

ical relatives of adopted schizophrenics than among the biological relatives of adopted controls. What might account for such a finding?

The most obvious possibility is that of selective placement, a universal phenomenon in the real world in which adoptions in fact occur, and a phenomenon that undermines the theoretical separation of genetic and environmental variables claimed for adoption studies. The children placed into homes by adoption agencies are never placed randomly. For example, it is well known that biological children of college-educated mothers, when put up for adoption, are placed selectively into the homes of adoptive parents with higher socioeconomic and educational status. The biological children of mothers who are grade-school dropouts are usually placed into much lower status adoptive homes. Thus it seems reasonable to ask: Into what kinds of adoptive homes are infants born into families shattered by alcoholism, criminality, and syphilitic psychosis likely to be placed? Further, might not the adoptive environment into which such children are placed cause them to develop schizophrenia?

From raw data kindly made available to one of us by Dr. Kety, we have been able to demonstrate a clear selective placement effect. Whenever a record of psychiatric treatment of a relative was located by Kety's team, notation was made about whether the relative had been in a mental hospital, in the psychiatric department of a general hospital, or in some other facility. When we check the adoptive families of the schizophrenic adoptees, we discover that in eight of the families (24 percent) an adoptive parent had been in a mental hospital. That was not true of a single adoptive parent of a control adoptee. That, of course, is a statistically significant difference—and it suggests as a credible interpretation of the Kety et al. results that the schizophrenic adoptees, who indeed had been born into shattered and disreputable families, acquired their schizophrenia as a result of the poor adoptive environments into which they were placed. The fact that one's adoptive parent goes into a mental hospital clearly does not bode well for the psychological health of the environment in which one is reared. There is, by the way, no indication that the biological parents of the schizophrenic adoptees have been in mental hospitals at an excessive rate. That occurred in only two families (6 percent), a rate in fact lower than that observed in the biological families of the control adoptees.

The same set of subjects has also been reported on in a later paper by Kety et al.[39] For this later work as many as possible of the relatives of index and control adoptees had been traced down personally and interviewed by a psychiatrist. The interviews were edited, and consensus diagnoses were then made blindly by the investigators. The basic picture did not change much. There were more spectrum diagnoses among relatives of index cases than among relatives of controls, although the interview procedure greatly increased the overall frequency of such diagnoses. This time, however, diagnoses of inadequate personality had to be excluded from the spectrum, since they occurred with equal frequency in both sets of relatives. The significance of the 1968 results, based on records rather than interviews, had depended upon including inadequate personality in the elastic spectrum.

Personal correspondence with the psychiatrist who conducted the interviews with relatives has revealed a few interesting details. The 1975 paper speaks only of "interviews," but it turns out that in several cases, when relatives were dead or unavailable, the psychiatrist "prepared a so-called pseudo interview from the existing hospital records." That is, the psychiatrist filled out the interview form in the way in which he guessed the relative would have answered. These pseudo interviews were sometimes diagnosed with remarkable sensitivity by the team of American investigators. The case of the biological mother of S-ii, a schizophrenic adoptee, is one particularly instructive example.

The woman's mental hospital records had been edited and then diagnosed blindly by the investigators in 1968. The diagnosis was inadequate personality—at that time, inside the spectrum. The 1975 paper—by which time inadequate personality is outside the spectrum —indicates that, upon personal interview, the woman had been diagnosed as a case of uncertain borderline schizophrenia—again inside the spectrum. But personal correspondence has revealed that the woman was never in fact interviewed; she had committed suicide long before the psychiatrist attempted to locate her, and so—from the original hospital records—she was "pseudo interviewed." Perhaps the most remarkable aspect of the story, also revealed by personal correspondence, is that the woman had been hospitalized twice—and each time had been diagnosed as manic-depressive by the psychiatrists who actually saw and treated her. That is, she had been diagnosed as suffering from a mental illness unrelated to schizophrenia, and very clearly outside the schizophrenia spectrum. We can only marvel at the fact that the American diagnosticians, analyzing abstracts of these same records, were twice able to detect—without ever seeing her—that she really belonged within the shifting boundaries of the spectrum.

The Kety study has more recently been expanded to include all of Denmark (rather than merely Copenhagen). The hospital records of relatives have been searched and the results briefly referred to in a couple of publications. The relatives are also being interviewed. There have been no detailed data published or made available for the larger sample, so critical analysis is not yet possible. Though Kety asserts that results from the expanded sample confirm those earlier reported in detail, there is no reason to suppose that the more recent work is free of the invalidating flaws we have outlined above.

These results must be evaluated together with the results of a companion study reported by Rosenthal et al. using the same Danish files.[40] This study first identified a number of schizophrenic parents who had given up children for adoption. The question is whether those children, not reared by their schizophrenic biological parents, will tend to develop schizophrenia. The control group for the index children was made up of adoptees whose biological parents had no record of psychiatric treatment. The index adoptees and the controls, when grown up, were interviewed—blindly—by a Danish psychiatrist. Based upon those interviews, decisions were made as to whether particular individuals were in or out of the spectrum of schizophrenic disorders. Countless textbooks now indicate that a higher frequency of spectrum disorders were diagnosed in the adopted children of schizophrenics than in children of normal controls. That claim is based on preliminary (and inadequately reported) accounts of the study.

The preliminary reports did claim to observe a barely significant tendency for spectrum disorders to be more frequent among the index cases. (There was only one adoptee who had ever in fact been hospitalized for schizophrenia, and the authors frankly admitted that if they had looked only for hospitalized cases of schizophrenia, "we would have concluded that heredity did not contribute significantly to schizophrenia.")[41] The early papers, however, are entirely vague as to when and how or by whom decisions were made about whether individual cases were in or out of the spectrum. The papers indicate merely that the interviewing Danish psychiatrist made a "thumbnail diagnostic formulation" for each interview, and that these were somehow related to whether or not the interviewee was placed into the spectrum. Personal correspondence with several of the collaborators has made it clear that the "thumbnail diagnostic formulation" of the interviewer did not specify whether the individual was in or out of the spectrum. For the early papers, that decision was made in a manner and by parties unknown.

When consensus diagnoses like those in the Kety study were reported on for the first time in 1978, it developed that there was no significant tendency for spectrum cases to occur more frequently among index subjects.[42] Thus, despite the widely cited misleading early reports of the Rosenthal et al. study, its outcome was in fact negative.

Wender et al. added a new refinement to the Rosenthal study by reporting on a new group of twenty-eight "cross-fostered" subjects.[43] These were adoptees whose biological parents had been normal but whose adoptive parents had become schizophrenic. The new group was added to observe whether the experience of being reared by a schizophrenic adoptive parent would produce pathology in a child. The cross-fostered children, according to Wender et al., did not show more pathology than did the control adoptees. But it is important to note that in this paper the concept of diagnosing a schizophrenia spectrum has been abandoned; instead, the Danish interviews were now being rated for "global psychopathology." Consensus diagnoses —or any other diagnoses—of whether or not the cross-fostered children were in the schizophrenia spectrum have not appeared in any of the many papers concerned with the genetics of schizophrenia.

There is, however, an obscure paper from the Kety and Rosenthal group concerned with the characteristics of people who refuse to take part in psychological studies that contains some important and relevant information.[44] The paper includes as an aside an incidental table (Table 14) showing the percentage of spectrum diagnoses made in each group by a Danish psychiatrist, Schulsinger. We learn from that table that fully 26 percent of the cross-fostered adoptees were diagnosed as being in the schizophrenia spectrum—a rate not significantly different from that of the index adoptees themselves. Further, that obscure table is the only place where data on an immensely relevant control group have been reported. The Danish investigators, it turns out, also interviewed (and diagnosed) a number of nonadopted children of schizophrenics, who had been reared by their mentally ill biological parents. The rate of spectrum disorder among this group did not differ from that observed among cross-fostered children. Thus, had they taken the design of their own study seriously, the investigators might have concluded that they had shown schizophrenia to be entirely of environmental origin. The cross-fostered biological children of normal parents, when merely reared by schizophrenic adoptive parents, show just as great a frequency of spectrum disorders as do the nonadopted biological children of schizophrenics. The reader may not be surprised to learn that consensus diagnoses of the nonadopted group, like consensus diagnoses of the cross-fostered group, have never been reported.

The weaknesses of the Danish adoption studies are so obvious upon critical review that it may be difficult to understand how distinguished scientists could have regarded them as eliminating all the artifacts that beset family and twin studies of nature and nurture. In fact, a team of investigators from the French National Institute of Medical Research have published, quite independently, an analysis of the Danish adoption studies that reaches the conclusion that they are gravely deficient.[45] Perhaps one factor encouraging the usually uncritical acceptance of the investigators' claims has been indicated by Wender and Klein in an article written for the popular magazine *Psychology Today*.[46] They cite the Danish adoption study—based upon a broad concept of schizophrenia spectrum—as indicating that "for each schizophrenic there may be 10 times as many people who have a milder form of the disorder that is genetically . . . related to the most severe form . . . 8 percent of Americans have a lifelong form of personality disorder that is genetically produced. This finding is extremely important." The importance of the finding is spelled out by Wender and Klein in the following language: "The public is largely unaware that different sorts of emotional illnesses are now responsive to specific medications and, unfortunately, many doctors are similarly unaware." The logic, erroneous at every step, is as follows: The Danish adoption studies have shown that schizophrenia, and a number of behavioral eccentricities, are genetically produced. Since the genes influence biological mechanisms, it must follow that the most effective treatment for schizophrenia, and for behavioral eccentricity, is drug treatment. Focusing on social or environmental conditions as a cause of disordered behavior would be fruitless.

Yet any materialist understanding of the relationship of brain to behavior must recognize that even if schizophrenia were largely genetic in origin, it would in no way follow that drugs—or any biological, as opposed to social, treatment—would necessarily be the most effective therapy. Just as drugs change behavior, so will altered behavior imposed by talking therapies change brains (as indeed the latent theory behind behavior modification would itself agree). The logic of this does not depend on a belief in any more explicit integration of the biological and the social.

Schizophrenia as Socially Determined

To reveal, as we have tried, the theoretical and empirical impoverishment of the conventional wisdom of biological determinism in relationship to schizophrenia does not then argue that there is nothing relevant to be said about the biology of the disorder, and still less does it deny that schizophrenia exists. The problem of understanding the etiology of schizophrenia and a rational investigation of its treatment and prevention is made vastly more difficult, perhaps even hopelessly tangled, by the extraordinary latitude and naiveté of diagnostic criteria. Certainly one may wonder about the relevance of biology to the diagnosis of schizophrenia either by the forensic psychiatrists of the Soviet Union or by the British psychiatrist who diagnoses a young black as schizophrenic on the basis of his use of the religious language of Rastafarianism.[47]

Misgivings are not eased when one recalls a well-known study by Rosenhan and his colleagues in California in 1973.[48] Rosenhan's group of experimenters presented themselves individually at mental hospitals complaining of hearing voices. Many were hospitalized. Once inside the hospital, according to the strategy of the experiment, they declared that their symptoms had ceased. However, it did not prove so easy to achieve release. The experimenters' claims to normality were disregarded, and most found themselves treated as mere objects by nurses and doctors and released only after considerable periods of time. A pseudo-patient who took notes in one of the hospitals, for instance, was described by nurses as showing "compulsive writing behavior."

Even more revealing, perhaps, was the drop in hospital admissions for schizophrenia in the area after Rosenhan circulated the results of the first experiment among doctors and indicated that they might be visited by further pseudo-patients in the future, although none were actually sent.

It is this sort of experience that lies behind the argument, developed in its most extreme form by Michel Foucault and his school over the last two decades, that the entire category of psychological disorders is to be seen as a historical invention, an expression of power relationships within society manifested within particular families. To simplify Foucault's intricate argument, he claims that all societies require a category of individuals who can be dominated or scapegoated, and over the centuries since the rise of science—and particularly since the industrial revolution of the nineteenth century—the mad have come to fill this category. In medieval times, he says, houses of confinement were built for lepers, and madness was often explained in terms of possession by demons or spirits.[49] According to Foucault the idea of institutionalizing the mad developed during the eighteenth and nineteenth centuries after the clearing of the leper houses left a gap for new scapegoats to replace the old ones.

In this view madness is a matter of labeling; it is not a property of the individual but merely a social definition wished by society on a proportion of its population. To look for correlates of madness in the brain or the genes is therefore a meaningless task, for it is not located in the brain or the individual at all. To dismiss the suffering and the deranged behavior of the schizophrenic merely as a problem of social labeling by those who have power over those who have not seems a quite inadequate response to a complex social and medical problem. Despite Foucault's historiography and the enthusiasm of its reception in Britain and France at the crest of the wave of antipsychiatry of the 1960s and 1970s, the actual historical account he gives of when and how asylums for the insane arose has been called into question.[50] And by cutting the phenomenon of schizophrenia completely away from biology and locating it entirely in the social world of labeling, Foucault and his followers arrive, from a very different starting point, back in the dualist Cartesian camp, which, as we showed in Chapters 2 and 3, preceded the full-blown materialism of the nineteenth century. So much has Foucault retreated that at certain points in his argument he even seems to be ambiguous as to whether "physical" quite apart from "mental" illness exists except in the social context that proclaims it.

More modest than Foucault's grand theorizing but nonetheless culturally determinist are the social and familial theories of schizophrenia developed by R. D. Laing.[51] For Laing—at least the Laing of the sixties and early seventies—schizophrenia is essentially a family disorder, not a product of a sick individual but of the interactions of the members of a sick family. Within this family, locked together by the nuclear style of living of contemporary society, one particular child comes to be picked upon, always at fault, never able to live up to parental demands or expectations. Thus the child is in what Laing calls (in a term derived from Gregory Bateson) a double bind; whatever he or she does is wrong. Under such circumstances the retreat into a world of private fantasy becomes the only logical response to the intolerable pressures of existence. Schizophrenia is thus a rational, adaptive response of individuals to the constraints of their life. Treatment of the schizophrenic by hospitalization or by drugs is therefore not seen as liberation from the disease but as part of that person's oppression.

Family context may be crucial in the development of mental illnesses such as schizophrenia, but it is clear that a larger social context is also involved. The diagnosis is made most often of working-class, inner-city dwellers, least often of middle- and upper-class suburban dwellers.[52] To a social theorist, the argument about the social context that determines the diagnosis is clear. An example of the class nature of the diagnosis of mental illness comes from the studies of depression by Brown and Harris in 1978 in Camberwell, an inner-city, largely working-class area of London, with some pockets of middle-class infiltration.[53] They showed that about a quarter of working-class women with children living in Camberwell were suffering from what they defined as a definite neurosis, mainly severe depression, whereas the incidence among comparable middle-class women was only some 6 percent. A large proportion of these depressed individuals, who if they had attended psychiatric clinics would have been diagnosed as ill and medicalized or hospitalized, had suffered severe threatening events in their lives within the past year, such as loss of husband or economic insecurity. The use of drugs—mainly tranquilizers—among such groups of women is clearly very high.

Biological determinism faces such social evidence with arguments that, for example, people with genotypes predisposing toward schizophrenia may drift downward in occupation and living accommodation until they find a niche most suited to their genotype. But it would be a brave biological determinist who would want to argue that in the case of the depressed housewives of Camberwell it was their genes that were at fault.*

An adequate theory of schizophrenia must understand what it is about the social and cultural environment that pushes some categories of people toward manifesting schizophrenic symptoms; it must understand that such cultural and social environments themselves profoundly affect the biology of the individuals concerned and that some of these biological changes, if we could measure them, might be the reflections or correspondents of that schizophrenia with the brain. It may well be that, in our present society, people with certain genotypes are more likely than others to suffer from schizophrenia—although the evidence is at present entirely inadequate to allow one to come to that conclusion. This says nothing about the future of "schizophrenia" in a different type of society, nor does it help us build a theory of schizophrenia in the present. Neither biological nor cultural determinism, nor some sort of dualistic agnosticism, is adequate to the task of developing such a theory. For that, we must look to a more dialectical understanding of the relationship between the biological and the social.

*Brave but not impossible. In 1979 B. L. Reid and his colleagues published a paper in the *Australian Medical Journal* claiming that the higher incidence of uterine cancer among working-class women was due to a factor carried in the sperm of their working-class male partners, and that the same working-class sperm had a simpler, more repetitive structure to its DNA than did that of middle-class sperm. This accounted for working-class people only being able to think simple and repetitive thoughts, unlike the complexity available to the middle classes.[54] To such preformationist thinking, clearly, no sort of biological determinism is impossible.

CHAPTER NINE

SOCIOBIOLOGY: THE TOTAL SYNTHESIS

In the spring of 1975 a remarkable event in academic publishing took place. Harvard University Press, using the full panoply of public relations devices—including full-page advertisements in the *New York Times,* author-publisher cocktail parties, prepublication reviews and interviews on television, radio, and in popular magazines[1]—issued a book on evolutionary theory by an expert on ants. While evolutionary theory 116 years after Darwin's *Origin of Species* would hardly seem to be startling enough to warrant great public excitement, and professors of zoology are not often the subject of interviews in household magazines, the book *Sociobiology: The New Synthesis*[2] and its author, E. O. Wilson, soon attained considerable celebrity. Clearly the publishers expected and promoted the book's popularity, both by their publicity campaign and by the coffee-table format of the work itself, large and lavishly illustrated with original drawings of animal societies. But even so, a 600-page book filled with such subjects as mathematical population genetics, neurobiology, and primate taxonomy, requiring an extensive glossary to be accessible to its readers, does not often make the pages of *House and Garden, Readers Digest,* and *People* magazine.[3] Nor does it often, at $25.00, sell over 100,000 copies. What gave *Sociobiology* its immense interest outside of biology was the extraordinary breadth of its claims. In the introductory chapter, entitled "The Morality of the Gene," Wilson defines sociobiology as "the systematic study of the biological basis of all social behavior. For the present it focuses on animal societies. . . . But the discipline is also concerned with social behavior of early man and the adaptive features of organization in the more primitive human societies." The book as a whole was intended to "codify sociobiology into a branch of evolutionary biology" encompassing all human societies, ancient and modern, preliterate and post-industrial. Nothing is left out, since "sociology and the other social sciences, as well as the humanities, are the last branches of biology waiting to be included in the Modern Synthesis. One of the functions of sociobiology, then, is to reformulate the foundations of the social sciences in a way that draws these subjects into the Modern Synthesis"(page 4).

In the book that follows, the author offers a biological explanation of such human cultural manifestations as religion, ethics, tribalism, warfare, genocide, cooperation, competition, entrepreneurship, conformity, indoctrinability, and spite (this list is incomplete). Wilson, however, is not content merely to explain the world. The point is to change it. Starting from a program to understand all of society, he ends with a vision of neurobiologists and sociobiologists as the technocrats of the near future who will provide the necessary knowledge for ethical and political decisions in the planned society:

If the decision is taken to mold cultures to fit the requirements of the ecological steady state, some behaviors can be altered experientially without emotional damage or loss in creativity. Others cannot. Uncertainty in the matter means that Skinner's dream of a culture predesigned for happiness will surely have to wait for the new neurobiology. A genetically accurate and hence completely fair code of ethics must also wait. (Page 575)*

*B. F. Skinner, the behaviorist psychologist, believes that human beings can be programmed by early conditioning to behave in predetermined ways, including the possibility of conditioning them for a utopian society. See, for example, his *Beyond Freedom and Dignity* and *Walden II.* (See also Chapter 6.)

We must wait for sociobiologists to provide the scientific tools of correct social organization because

we do not know how many of the most valued qualities are linked genetically to the more obsolete, destructive ones. Cooperativeness toward groupmates might be coupled with aggressivity toward strangers, creativeness with a desire to own and dominate, athletic zeal with a tendency to violent response, and so on. . . . If the planned society—the creation of which seems inevitable in the coming century—were to deliberately steer its members past those stresses and conflicts that once gave the destructive phenotypes their Darwinian edge, the other phenotypes might dwindle with them. In this, the ultimate genetic sense, social control would rob man of his humanity. (Page 575)

Not since Hobbes's *Leviathan* has there been such an ambitious program to explain and prescribe for the entire human condition beginning with a few basic principles. But unlike Hobbes, Wilson is not a children's tutor whose only authority is the weight of his own argument. He speaks with the voice of modern biology, that most prestigious of sciences. Professional biologists and anthropologists seized on sociobiology as quickly as the popular press. Following the publication of Wilson's book, a stream of works echoing, modifying, and extending the theme of sociobiology rapidly appeared,[4] and Wilson himself devoted a later work, *On Human Nature*, entirely to the question of human sociobiology.[5] There was, at least at first, virtually unanimous praise by biologists, who quickly recognized sociobiology as an official subdiscipline of evolutionary biology and anthropology.[6] Since 1975 at least three new scientific journals devoted to sociobiology have been started, edited collections of papers on sociobiology are common,[7] and scores of teaching and research positions in American and British universities have been created for sociobiologists at a time of shrinking budgets. Sociobiological explanations began to appear in the literature of economics and political science,[8] and *Business Week* offered "A Genetic Defense of the Free Market."[9]

The claim of sociobiology to explain all of the human condition may account for the initial interest in it, but not for the sympathy with which it has been greeted in the public media, nor its continued popularity as a paradigm in academic theory. It is the nature of the explanation itself that has had such immense appeal. The central assertion of sociobiology is that all aspects of human culture and behavior, like the behavior of all animals, are coded in the genes and have been molded by natural selection. While sociobiologists sometimes hedge on the issue of direct genetic determination of every detail of social and individual behavior, the claim for ultimate genetic control, as we shall see, lies at the heart of a system of explanation that cannot survive otherwise. Although sociobiologists, when challenged by geneticists, at times retreat to the position that they are only claiming that genes determine the possible range of human behaviors, sociobiology is emphatically *not* simply the claim that human society is of a nature made possible by human biology. All manifestations of human culture are the result of the activity of living beings; therefore it follows that everything that has ever been done by our species individually or collectively must be biologically possible. But that says nothing except that what has actually happened must have been in the realm of possibility. Whatever it is, sociobiology is not a simple tautology.

Sociobiology is a reductionist, biological determinist explanation of human existence. Its adherents claim, first, that the details of present and past social arrangements are the inevitable manifestations of the specific action of genes. Second, they argue that the particular genes that lie at the basis of human society have been selected in evolution because the traits they determine result in higher reproductive fitness of the individuals that carry them. The academic and popular appeal of sociobiology flows directly from its simple reductionist program and its claim that human society as we know it is both inevitable and the result of an adaptive process.

The general appeal of sociobiology is in its legitimation of the status quo. If present social arrangements are the ineluctable consequences of the human genotype, then nothing of any significance can be changed. So, Wilson predicts that

the genetic bias is intense enough to cause a substantial division of labor even in the most free and most egalitarian of future societies. . . . Even with identical education and equal access to all professions, men are likely to continue to play a disproportionate role in political life, business and science.[10]

What is not always realized is that if one accepts biological determination, nothing need be changed, for what falls in the realm of necessity falls outside the realm of justice. The issue of justice arises only when there is choice. Sociobiologists are not consistent on this point. In *Sociobiology,* Wilson committed the naturalistic fallacy of "the genetically accurate and hence completely fair code of ethics," but shortly after, in *Human Decency Is Animal* he cautioned against deriving "ought" from "is." The effective political truth, however, is that "is" abolishes "ought." To the extent that we are free to make ethical decisions that can be translated into practice, biology is irrelevant; to the extent that we are bound by our biology, ethical judgments are irrelevant. It is precisely because biological determinism is exculpatory that it has such wide appeal. If men dominate women, it is because they must. If employers exploit their workers, it is because evolution has built into us the genes for entrepreneurial activity. If we kill each other in war, it is the force of our genes for territoriality, xenophobia, tribalism, and aggression. Such a theory can become a powerful weapon in the hands of ideologues who protect an embattled social organization by "a genetic defense of the free market." It also serves at the personal level to explain individual acts of oppression and to protect the oppressors against the demands of the oppressed. It is "why we do what we do"[11] and "why we sometimes behave like cavemen."[12]

The claim that genetically determined social organization is a product of natural selection has the further consequence of suggesting that society is in some sense optimal or adaptive. While genetic fixity in itself is logically quite sufficient to support the status quo, the claim that present social arrangements are also optimal adds to their palatability. It is rather a handy feature of life that what must be is also the best. In Voltaire's *Candide,* the philosopher, Dr. Pangloss, insists that this is the "best of all possible worlds." Sociobiology is Pangloss made scientific through the agency of Charles Darwin. This convergence of possible and optimal has long been a characteristic argument in favor of capitalism; those who promote such a view claim it is the only possible mode of economic organization in a world of scarce resources and greedy people, and they sometimes argue that it is the most efficient organization of production and distribution. There are deep contradictions within sociobiology on the issue of optimality and adaptation. On the one hand, the technical argument of sociobiology specifically rejects benefit to the individual, group, or species as a motive force in evolution and places entire reliance on the mechanical consequences of differential reproduction of genotypes. Indeed, what distinguishes modern sociobiology from older attempts to explain the evolution of behavior is its explicit rejection of the selection of entire groups and its concentration on the gene as the unit of natural selection. Even the individual may not benefit, only the gene. In its vulgarized form, it is the metaphor of the "selfish gene" of which "we are the survival machines—robot vehicles programmed to preserve the selfish molecules known as genes."[13] On the other hand, sociobiologists use optimality arguments to derive their explanations and predictions. Many of these are derived from economic theory and concern the optimal use of time or energy by individuals or groups. Organisms are regarded as problem solvers who choose strategies for the optimal solution to environmental problems. While in principle such arguments could be framed entirely in terms of the rate of reproduction of genes, in practice optimality arguments are substitutes for the rigidly mechanical calculus of gene reproduction. In fact, optimality arguments lie at the heart of the sociobiological method.

In addition to the political appeal of sociobiology as legitimating a hierarchical, entrepreneurial, competitive society, it has a strong attraction for bourgeois intellectuals because of its extreme reductionism. Anthropologists, sociologists, economists, and political scientists have no agreed-upon central body of theory. On the contrary, there are competing modes of explanation for the same phenomena. The record of successful prediction and manipulation of the real world of economics and politics is dismal. At the same time, many of those who study social phenomena have been attempting to assimilate themselves into the world of natural sciences, calling themselves "social scientists" and using those accoutrements of natural science, statistics and mathematics, to become more exact. The promised biologization of social studies is precisely a realization of the desire of sociologists, anthropologists, and economists to be *scientists.* Moreover, the simple calculus of genetic advantage is a speculative game that anyone can play. Into the sterile desert of sociological controversy has flowed the fertilizing stream of biological explanation, and a hundred flowers have bloomed. From the complete system of capitalist production and distribution,[14] to ethics and moralizing,[15] through the Kent State massacre,[16] Soviet military intentions,[17] and the alleged preference of the upper middle class for cunnilingus and fellatio,[18] everything is explained as the product of selected genes. Minds starved for something new to express have found their sustenance. At the same time, long-standing conflicts between reductionists and nonreductionists have been intensified so that some of the most penetrating and scathing critiques of sociobiology have come from anthropologists and social philosophers.[19] The intellectual imperialism of a new discipline that threatens to engulf all other intellectual domains cannot help but galvanize the long-slumbering resentment of students of society against the hubris of natural scientists. In doing so it intensifies the contradictions between the reductionist trend of bourgeois thought and the evident failure of reductionism as a methodological program in the study of society.

The Origins of Sociobiology

The appearance in 1975 of Wilson's manifesto was only one stage in the development of sociobiology. Its most immediate predecessors were a group of works on human nature that Stephen Gould has aptly characterized as "pop ethology": Robert Ardrey's *The Territorial Imperative* (1966); Konrad Lorenz's *On Aggression* (1966); Desmond Morris's *The Naked Ape* (1967); and Tiger and Fox's *The Imperial Animal* (1970). These books take the view that human beings are by nature territorial and aggressive. The human condition, for them, is Hobbes's war of all against all, a condition they derive from the fragmentary and controversial evidence of human paleontology and animal behavior. Ardrey, for example, based his argument on the supposition that *Homo sapiens* is descended from a nasty carnivorous hominid, *Australopithecus africanus*, which hunted down and extinguished its larger, more placid vegetarian relative, *Australopithecus robustus*. The argument, however, is fallacious. The claim that *africanus* was carnivorous is based on Ardrey's misunderstanding of the relatively larger canine teeth in this species. In primate evolution, teeth have grown larger more slowly than body size, so smaller apes always have relatively larger teeth, irrespective of diet. In fact, *africanus* and *robustus* have teeth in precisely the proportions for primates of their sizes.[20] The evidence that *africanus* was the ancestor of *Homo sapiens* has evaporated with the discovery that the already human toolmaker *Homo habilis* was contemporaneous with it. Ironically, Lorenz's claim for the innate nastiness of humans is the reverse of Ardrey's. He says we come from vegetarian ancestors who, lacking the sharp teeth and other natural weapons of a predator, also lack the built-in behavioral avoidance of mortal combat that protects predators from destroying each other. In either case, the evidence has clearly been sifted and selected to support the a priori view of an innately aggressive, territorial, entrepreneurial, male-dominated species. The political implications are clear and explicit. A fair sample is Ardrey's assertion that patriotism and private property are innate:

If we defend the title to our land or the sovereignty of our country, we do it for reasons no different, no less innate, no less ineradicable, than do lower animals. The dog barking at you from behind his master's fence acts for a motive indistinguishable from that of his master when the fence was built.[21]

While Wilson, in *Sociobiology*, wisely sought to distance himself from pop ethologies by calling them "works of advocacy,"[22] there seems not much to choose between Ardrey's simplistic generalization and such insights into human nature as "man would rather believe than know"[23] or "human beings are absurdly easy to indoctrinate—they seek it"[24] in which *Sociobiology* abounds.

Sociobiology and the pop ethologies are forms of human nature theory that in some aspect characterize all political philosophy. Every theory of society implies a theory of what it is to be human. Every theorist of society carries out the same fiction of apparently deducing the nature of society from a priori considerations of the innate nature of human beings, while in fact inducing the necessary assumptions from the end to be reached. In hypostasizing entrepreneurial bourgeois society, sociobiology is a direct intellectual descendent of Thomas Hobbes's *Leviathan* of 1651.[25] Hobbes explicitly modeled his argument on Galileo's method of reduction and recomposition of a system. He first resolved society into its elements, individual human beings, and then further reduced them to individual elements of motion. Humans were automated machines whose operation led ineluctably to certain social phenomena. The competitive behavior of human beings in society was not, for Hobbes, a primary innate feature, but was a consequence of the *social* life of machine-organisms attempting to maintain themselves in a world of finite resource. In this sense Hobbes was both more reductionist and yet more sophisticated than sociobiologists. He postulated many fewer basic instinctual elements in human nature from which all else was derived, but in so doing he recognized that social interaction was the necessary condition for the occurrence of competition. The war of all against all was the rational and prudent behavior of the human machine when in society. As Macpherson has clearly shown,[26] the logic of the argument required that Hobbes had in mind bourgeois society in which individuals' labor power is their own property and, together with all other forms of property, is alienable. Thus, Hobbes's political theory is a classic of seventeenth-century thought combining, as we described in Chapters 3 and 4, the extreme reductionism of the new bourgeois science with the individualism and alienability of property of bourgeois productive relations.

The influence of Hobbes's thought on sociobiology comes not directly but through the intermediary of Darwinism and social Darwinism. It is common to characterize Darwinism as "Hobbesian" because of its emphasis on the struggle for existence, but the similarity is both deeper and more ambiguous. Competition for Darwin, as for Hobbes, was not a fundamental property of organisms but the consequence of the automatic self-reproduction of the machine-organism in a world of finite resource. This enabled Darwin to understand the struggle for existence in a very broad sense depending upon the particular interaction between organism and environment. Thus he writes of the struggle for existence:

I should premise that I use the term in a large and metaphorical sense including dependence of one being on another, and including (which is more important) not only the life of the individual, but success in leaving progeny. Two canine animals, in a time of dearth, may truly be said to struggle with each other which shall get food and live. But a plant at the edge of the desert is said to struggle for life against the drought.[27]

It was this mention of dependence of one being on another by Darwin, and his discussion of some cases in *The Descent of Man*, that allowed Kropotkin to identify himself as a Darwinian in his own emphasis on cooperation.[28] Yet there is no doubt that, as Kropotkin ruefully observed, Darwin himself and most of his followers emphasized the competitive struggle between organisms. This should not surprise us. That the Hobbesian element dominates Darwin's thought is evidence both of the Malthusian origin of the *Origin* and of the pervasiveness of competitive relations in our society. Darwin transferred the idea of competition from society to biology. Spencer had already coined the term "survival of the fittest" in *Social Statics* of 1862, and the social Darwinism of the later nineteenth century might better be called "Spencerism."[29] The justification for laissez-faire capitalism on the basis of Darwinian theory only completed a historical circle.[30]

Thus, Darwinism came to be used throughout the late nineteenth and early twentieth centuries to reinforce, by a secondary derivation, the Hobbesian-Malthusian-Spencerian view that society progressed by survival of the fittest in a competitive struggle. Entrepreneurial activity, the subjection of one group to another, the subjection of "lower races" were all seen to be both part of human nature and, at the same time, part of a universal law of survival. Andrew Carnegie assured the readers of the *North American Review* that "it is here; we cannot evade it; no substitutes for it have been found; and while the law may be hard for the individual, it is best for the race, because it assures survival of the fittest in every department"[31]—including, presumably, the steel-manufacturing department. War and conquest, too, were laws of nature:

The greatest authority of all of the advocates of war is Darwin. Since the theory of evolution has been promulgated, they can cover their natural barbarism with the name of Darwin and proclaim the sanguinary instincts of their inmost hearts as the last word of science.[32]

A direct line connects this tradition to Wilson's assertion that "the most distinctive human qualities" emerged during the "autocatalytic" phase of social evolution that occurred through "intertribal warfare," "genocide," and "genosorption"[33] (the merging of the genes of the conquered with the conquerors).

The principles of Darwinism could also be used to make an all-encompassing theory of society. The arch social Darwinist, William Graham Sumner, found in 1872 that the struggle for existence "solved the old difficulty about the relations of social science to history, rescued social science from the dominion of the cranks, and offered a definite and magnificent field to work, from which we might hope at last to derive definite results for the solution of social problems."[34] The New Synthesis is not so new after all. There is really nothing that separates the program or specific claims of the social Darwinism of the 1870s from the Darwinian sociobiology of the 1970s.

The embarrassment of the bar sinister in the intellectual ancestry of sociobiology has led many biologists who work in the sociobiological mode to disclaim the specifically human implications of their work. For them, sociobiology is simply the study of the evolution of social behavior in all sorts of animals that do not have the unpleasant complication of culture and abstract thought. Indeed, when *Sociobiology* was first attacked for presenting political conclusions about human society in the guise of evolutionary theory,[35] many biologists took the charitable view that the material on humans in that book was added as an afterthought to add interest to what was otherwise a heavy academic tome. The development of the literature of sociobiology since 1975, however, including Wilson's own *On Human Nature*, leaves little doubt that the problem of human nature is at the center of sociobiological concerns. There may indeed be a field of sociobiology that is concerned with the evolution of animal behavior, although what distinguishes it from evolutionary biology in general and evolutionary ethology in particular is unclear. What does seem clear is that sociobiologists wish to have it both ways. They would like the notoriety associated with the name "sociobiology" because of the prosperity it has brought to a previously depressed sector of the intellectual economy, while rejecting (always quietly) the source of their wealth. "He that lies with the dogs, riseth with fleas."

The Argument of Sociobiology

Sociobiology, as a theory of human society, is built of three parts. First, there is a description of the phenomenon that it is meant to explain, that is, a statement of human nature. This description consists of an extensive list of characteristics that are thought to be universals in human societies, including such diverse phenomena as athletics, dancing, cooking, religion, territoriality, entrepreneurship, xenophobia, warfare, and the female orgasm.

Second, having described human nature, sociobiologists claim that the universal characteristics are coded for in the human genotype. There is, as we shall see, a great deal of confusion, imprecision, and internal contradiction as to what sociobiologists mean by genetic control and innateness, so that almost any statement about the relations between genes and culture can be supported with appropriate quotations. At times, direct genetic control of specific universals is projected, as for example postulated conformer genes[36] or genes for reciprocal altruism.[37] At others, it is stated only that "the genes hold culture on a [very long] leash."[38] At the very least, sociobiologists argue that the specific content of human social organization that is supposed to be universal is itself a consequence of gene action. It is not that the complex human central nervous system *allows* people to imagine gods, but the human genome *demands* that they do so.

The third step in the sociobiological argument is the attempt to establish that the genetically based human social universals have been established by natural selection during the course of human biological evolution. The method consists essentially of contemplating the trait and then making an imaginative reconstruction of human history that would have made the trait adaptive, or would have led the possessors of the hypothetical genes for the trait to leave more offspring.

In what follows, we look more closely into these three elements of sociobiology: the description of human nature, the claim of its innateness, and the argument for its adaptive origin.

THE PICTURE OF HUMAN NATURE

It seems only reasonable that those who see themselves as constructing a new science, hailed by many as revolutionary, would begin by a searching examination of their methodology of description. This is especially true when the data are historical, sociological, and anthropological. While we would hold that there are no "objective" or "scientific" descriptions of human social organization that go beyond the trivial, and that the goal of purging ideology from sociology is illusory, we can expect that students of human society will at least recognize the problem. Conventional social science has long done so and has sometimes tried to cope with the more obvious biases of ethnocentricity, sex, and political ideology. Yet the deep epistemological problems that face anyone who wishes to describe "human nature" seem not to have been taken into account by sociobiological theorists. Faced with the extraordinary richness of complexity of human social life in the past and the present, they have chosen the nineteenth-century path of describing the whole of humankind as a transformation of European bourgeois society. Wilson's description of human political economy is an example:

The members of human societies sometimes cooperate closely in insectan fashion, but more frequently they compete for the limited resources allocated to their role sector. The best and most entrepreneurial of the role-actors usually gain a disproportionate share of the rewards, while the least successful are displaced to other, less desirable positions.[39]

That this description of a possessive individualist entrepreneurial society would apply to the peasant economy of eleventh-century France or the serfs of Eastern Europe or Mayan and Aztec peasants seems patently wrong. And who are these abnormal insectan hordes of cooperators? Perhaps the Maoist Chinese who were "energized by the goals of collective self-aggrandizement."[40]

It would be difficult for anyone to present the entire set of social phenomena that are said to be human nature. Indeed, there is disagreement even among sociobiologists on an appropriate list. Roughly, humans are seen as self-aggrandizing, selfish animals whose social organization, even in its cooperative aspects, is a consequence of natural selection for traits that maximize reproductive fitness. In particular, humans are characterized by territoriality, tribalism, indoctrinability, blind faith, xenophobia, and a variety of manifestations of aggression. Unselfish behavior is really a form of selfishness in which the individual is motivated by an expectation of reciprocal reward. Self-righteousness, gratitude, and sympathy are examples, while aggressively moralistic behavior is a way of keeping cheaters in line. "Lives of the most towering heroism are paid out in the expectation of great reward." "Compassion . . . conforms to the best interests of self, family and allies of the moment." "No sustained form of human altruism is explicitly and totally self-annihilating."[41]

To universalize features of society through history and over cultures is not difficult. The very richness of the ethnographic record and the plasticity in its interpretations guarantee that large numbers of tribes said to display one phenomenon or another can be chosen as anecdotal cases. The amassing of supporting anecdotes is a standard method in works of advocacy. There are, however, cases that seem to contradict the claim of universality, but these can also be dealt with by standard techniques. One is the use of inclusive definition:

Anthropologists often discount territorial behavior as a general human attribute. This happens when the narrowest concept of the phenomenon is borrowed from zoology. . . . Each species is characterized by its own particular behavioral scale. In extreme cases the scale may run from open hostility . . . to oblique forms of advertisement or no territorial behavior at all. One seeks to characterize the behavioral scale of the species and to identify the parameters that move individual animals up and down it. If these qualifications are accepted, it is reasonable to conclude that territoriality is a general trait of hunter-gatherer societies.[42]

A second is to claim that the failure to display a universal trait is a temporary aberration. Although genocidal warfare is a supposed universal of human culture, "It is to be expected that some isolated cultures will escape the process for generations at a time, in effect reverting temporarily to what ethnographers classify as a pacific state."[43]

In our critique of sociobiology we will not attempt to argue for particular interpretations of the ethnographic record. We could do nothing more than engage in the same selective advocacy and reinterpretation that characterizes the work of sociobiologists. Detailed refutations of sociobiologists' interpretation of the literature of ethnography have been made by anthropologists,[44] but sociobiologists have their own coterie of sympathetic anthropologists.[45] The issue is not to decide whether Samoans are indeed pacific or aggressive but to understand how sociobiological descriptions allow an arbitrary interpretation of the record of human social organization that can be molded to fit the needs of the argument.

The problem that is nowhere faced is how to choose the universal characteristics of human nature in the face of immense individual and cultural variation. If aggression and patriotism are universal human traits, then was A. J. Mustie, who spent many years in jail for obstructing patriotic wars, other than human? On the other hand, if patriotic aggression is simply a variable part of the human repertoire, then in what sense except a trivial one is it more a part of human nature than, say, coprophilia? Indeed, the reader will be hard put to think of *any* behavior, no matter how bizarre, that has not been manifested by some number of people at some time.

The conventional description of human nature in sociobiological writing means that sociobiologists have failed to confront the fundamental problems of the description of behavior. They treat categories like slavery, entrepreneurship, dominance, aggression, tribalism, and territoriality as if they were natural objects having a concrete reality, rather than realizing that these are historically and ideologically conditioned constructs. Any theory of the evolution of, say, entrepreneurship depends critically on whether that concept has any reality outside the heads of modern historians and political economists. There are four specific sorts of error of description made by sociobiologists that deeply undermine any claim they have to illuminating human society.

First, sociobiology uses arbitrary agglomeration. One of the most difficult problems of description in evolutionary theory, and not sociobiology alone, is to decide how an organism is to be cut up into parts in understanding its evolution. What is the correct topology of description, the natural suture lines along which the phenotype of the individual is to be divided for the purposes of evolutionary theory? For example, is it illuminating to speak of the evolution of the hand? It might be that the hand is too small a unit and that only the evolution of the entire forelimb makes sense, or alternatively, that the separate fingers or even joints are the appropriate level of description. Indeed, paleontologists often speak of the evolution of the opposable thumb as of overwhelming importance in human history. There is no a priori way to decide on the appropriate level or levels of description. The answer depends in part upon the way in which the genes that influence the growth of the hand influence other aspects of development. But, also, changes in the hand alter the relation of the organism to the external world, which alteration in turn affects the pressure of natural selection on other aspects of the organism. That is, the hand is tied in evolution to other parts of the body both by internal and external relations. Until these are understood, it is by no means sure that the hand is an appropriate unit of phenotypic description.

An example of the critical importance of understanding developmental relations is the evolution of the chin. Human anatomical evolution can be described as neotenic, which means that, anatomically, human beings are like prematurely born apes. Human fetuses and ape fetuses are much more alike than are the adults, and the human adult resembles a fetal ape more than it does an adult ape. The single exception to this neotenic pattern is the human chin, which is more developed in the adult than in the fetus in humans but less in apes. Adaptive explanations of why the chin should be an exception to the general evolution of human shape could, with ingenuity, be constructed,[46] but the answer to the puzzle appears to be that the chin does not really exist as an evolutionary unit. There are two growth fields in the lower jaw: the dentary, which makes up the jaw bone itself, and the alveolar, which holds the teeth. Both of these have been undergoing the usual neotenic evolution in the human line, but the alveolar has shortened more rapidly than the dentary, with the result that a shape has evolved that we call the chin.

If it is difficult to decide how to divide up the anatomy of an organism for evolutionary explanation, how much more care must be used for behavior, especially in a social organism? It is already known that the topology of memory is not the same as the topology of the brain; specific memories are not stored in specific parts of the cerebral cortex but are somehow diffused spatially. Integrated cognitive function remains mysterious in its organization, yet sociobiologists find no problem at all in dividing up all of human culture into distinct evolving units.[47]

The second error of description is the confusion of metaphysical categories with concrete objects—the error of reification. As we have argued earlier in this book, it cannot be assumed that any behavior or institution to which a name can be given is a real thing subject to the laws of physical nature. Many of the mental objects that are said by sociobiologists to be units undergoing evolution are the abstract creations of particular cultures and times. What could "religion" have meant to the classical Greeks, who had no word for it and for whom it did not exist as a separate concept? Is "violence" real or is it a construct with no one-to-one correspondence with physical acts? What do we mean, for example, by "verbal violence" or a "violent exception"? The possession of real property as defined in modern law was unknown in thirteenth-century Europe, when the relationship was between persons rather than between a person and property that could be alienated. In fact, the relationship between a person and property which we call "ownership" is a legal fiction masking a social relation among persons that is only a few hundred years old in Europe.

Sociobiologists commit the classical error of reification by taking concepts that have been created as a way of ordering, understanding, and talking about human social experience and endowing these with a life of their own, able to act on the world and be acted upon. Just as the Greeks thought that those figments of the imagination, the gods, could reproduce and vanquish each other in battle, so sociobiologists think that religion can be inherited and increased in frequency by natural selection in the struggle to exist.

Third, metaphors are often taken for real identity, and the source of the metaphors is forgotten. There is a process of backward etymology in sociobiological theory in which human social institutions are laid on animals, metaphorically, and then the human behavior is rederived from the animals as if it were a special case of a general phenomenon that had been independently discovered in other species. A case that predates sociobiology but is incorporated into it is caste in insects. Caste is a human phenomenon, originally a race or lineage, but later a hereditary group associated with particular forms of labor and social position. By applying the idea of caste to insects, the sociobiologist gives legitimacy to the notion that human castes are simply an example of a more general phenomenon. But insects do not have castes, although they do indeed have individuals who are differentiated in their life activities. Indian castes were the result of the Aryan invasions and conquests of the Dravidian aborigines. High-caste Hindus had a monopoly of social and political power, while untouchables lived at the margins of existence. What has all this to do with ants? Does an ant queen (once called a king, before her sex was realized), a totally captive, force-fed, egg-bearing machine, have any resemblance to Elizabeth I or Catherine the Great, or even to the politically powerless but exceedingly rich Elizabeth II?

An illustration of the danger of these metaphors is in the phenomenon of "slavery" in ants. No modern sociobiologist derives human slavery biologically from ant slavery, and, as Wilson points out,[48] ant slavery, which involves the capture of one species by another, arose independently at least six times in the evolution of those insects. Yet language casts its magic spell. So Wilson writes:

The fact that slaves under great stress insist on behaving like human beings instead of slave ants, gibbons, mandrills or any other species is one of the reasons I believe that the trajectory of history can be plotted ahead, at least roughly. Biological constraints exist that define zones of improbable or forbidden entry.[49]

Slavery eventually fails in humans, according to this view, because the biological nature of humans causes them to resist the same institution that nonhumans suffer without struggle. The institution is general, the reaction to it specific. The view misses the point that "slavery" does not exist in ants. Slavery is a form of production of economic surplus, and slaves are a form of capital. Ants know neither commodities nor capital investment nor rates of interest nor the relative advantage to industrial capital of a free labor market.

While sociobiologists inherited royalty and slavery in ants from nineteenth-century entomology, they have made the false metaphor a device of their own. Aggression, warfare, cooperation, kinship, loyalty, coyness, rape, cheating, culture are all applied to nonhuman animals. Human manifestations then come to be seen as special, perhaps more developed, cases. Money is "a quantification of reciprocal altruism,"[50] and "the biological formula of territorialism translates easily into rituals of modern property ownership."[51]

A final problem of description, closely related to the use of metaphor, is the conflation of different phenomena under the same rubric. The classic that preoccupies sociobiologists and their predecessors is aggression. Originally meaning simply the unprovoked (but not necessarily irrational) attack of one person on another, aggression has come also to have a political meaning, the attack of one state on another, ultimately embodied in war. It is a reflection of the reductionist program of sociobiology that organized political aggression is seen as the collective manifestation of aggressive feelings of individuals against individuals, called into being by overcrowding and pressure for *Lebensraum*, or by the desire for mates. "Violent intergroup competition may occur over any scarce resource affecting reproductive success—land, animals, metals and so forth—but often it appears to occur over women, and even when women are not directly at issue, the combatants may recognize that women are at stake indirectly."[52]

Yet warfare among state-organized societies has little to do with prior individual feelings of aggression. War is a calculated political phenomenon undertaken at the behest of those in power in a society for political and economic gain. "Hostilities" begin without the least hostility between individuals except as deliberately created by the organs of propaganda. People kill each other in wars for all sorts of reasons, not the least of which is that they are forced to do so by the political power of the state. When the political power of the Russian state disintegrated in 1917, Russian soldiers stopped killing German soldiers. It is simply false that, as claimed by R. Trivers,[53] it is only necessary to play some martial music and men will march off to war impelled by their sexual instincts. Before the music come the schools and, if all else fails, the threat of prison or exile. Conflation is not simply a spontaneous error of unreflective sociobiological theorists. It is an essential step in the reductionist program.

THE INNATENESS OF BEHAVIOR

The central assertion of sociobiology is that human social behavior is, in some sense, coded in the genes. Yet as we have made clear in relation to IQ, up to the present time no one has ever been able to relate any aspect of human social behavior to any particular gene or set of genes, and no one has ever suggested an experimental plan for doing so. Thus, all statements about the genetic basis of human social traits are necessarily purely speculative, no matter how positive they seem to be.

What is it that sociobiologists assert about the relationship between descriptions of manifest traits like aggression and genes or human chromosomes? Sometimes a single gene coding for a given trait is postulated. Often the hypothetical nature of the gene is stated, but then the "if" is dropped from further discussion, treating the hypothetical model as real. In the process, the simple hypothetical gene may become a larger but unspecified number of genes. The relation between gene and trait is direct and determinative. Those who possess one form of the gene have the trait; those who carry a different form lack the trait or have it in smaller degree. Thus, Wilson writes of "societies contain-

ing higher frequencies of conformer genes,"[54] and "genetically programmed sexual and parent-offspring conflict."[55] One of the most
illuminating instances of the technique is the appearance of the mysterious "Dahlberg genes" for status. In *Sociobiology* we learn that "Dahlberg (1947) showed that if a single gene appears that is responsible for
success and an upward shift in status, it can be rapidly concentrated
in the uppermost socio-economic classes." Two paragraphs later we
are told that "there are many Dahlberg genes, not just the one postulated for argument in the simplest model."[56] The "if" has become
"there are." Still later in the development of sociobiological theory, the
"Dahlberg gene" is promoted to the status of a model of human cultural evolution, complete with a "method" and a "principal result."[57]
The citation of the well-known human geneticist Dahlberg may lead
the unwary reader to suppose that a serious scientific hypothesis was
being examined and a publishable result proved. In fact, the reference
is to a practice numerical problem at the end of a chapter in a textbook,[58] a game invented to help test a student's ability to manipulate
the algebra of genetics.

The trouble with the simple determinative model of gene control is
that the manifest traits of an organism, its phenotype, are not in general
determined by the genes in isolation but are a consequence of the
interaction of genes and environment in development. Sociobiologists
are aware of this fact, so sometimes they hedge. "It remains to be said
that if [homosexual] genes really exist, they are almost certainly incomplete in penetrance and variable in expressivity."[59] The trouble is that
if there are behavior genes in humans that affect only some unspecified
proportion of their carriers (incomplete penetrance) and with an unspecified variation in the nature of the effect (variable expressivity), no
geneticist can confirm their existence. The problem is exceedingly
difficult in experimental organisms where there is complete control
over environment and where experimental matings can be made. In
humans the problems of analysis are insuperable. When human genetics was in its primitive stages after Mendels' work was made known,
any trait whose inheritance was a completely impenetrable mystery
was passed off as a dominant gene with incomplete penetrance and
variable expressivity.

Sometimes sociobiologists say that the manifest trait is not itself
coded by genes, but that a potential is coded and the trait only arises

when the appropriate environmental cue is given. So Symons writes that "there is 'no aggressive drive' or accumulation of aggressive energy that must be discharged. . . . Natural selection favors willingness to fight only when benefits typically exceed costs in the currency of reproductive success, and in the absence of such circumstances, even a member of a typically aggressive species could live out its life span in peace."[60] Despite its superficial appearance of dependence on environment, this model is completely genetically determined, independent of the environment. The action of the genes is seen as creating a primeval computer program that will provide a fixed and stereotyped response to the appropriate signal. Of course, if the signal is never given, that part of the genetically determined central nervous circuitry is never activated.

Sometimes sociobiologists try to give both messages simultaneously: "Are human beings innately aggressive? . . . The answer . . . is yes. Throughout history, warfare . . . has been endemic to every form of society."[61] But as one reads on, it turns out that human aggressive behavior is "a structured, predictable pattern of interaction between genes and environment."[62] But we are on dangerous ground for sociobiology here. If aggression is manifest only in *some* environments, then in what important sense is it innate and why do we not simply avoid the wrong environments? It is at this point that notions alien to genetics begin to appear.

Human beings are strongly *predisposed* to respond with unnecessary hatred to external threats. . . . Our brains do appear to be programmed to the following extent: we are *inclined* to partition other people into friends and aliens. . . . We *tend* to fear deeply the actions of strangers. . . . The learning rules of violent aggression are largely obsolete. . . . But to acknowledge the obsolescence of the rules is not to banish them. We can only work our way around them. To let them rest latent and unsummoned, we must consciously undertake those difficult and rarely travelled pathways in psychological development that lead to mastery over and reduction of the profound human *tendency* to learn violence. [Emphases added throughout.][63]

What a thicket we must make our way through here! From the straightforward notion of behavior contingent on circumstance, we come to "tendencies," "predispositions," and "inclinations" to a behavior that is not dependent upon particular environments. Our brains are *programmed* to divide people into friends and strangers, and having done so, then to fear the strangers and, in the presence of that self-created threat, to respond violently. Despite the talk of the interaction of genes and environment, this is simply the theory that genes dictate aggressive behavior in social intercourse, but that overt aggression can be repressed by will or political structures.

A concept of gene action that permeates sociobiology is that alternative forms of social organization are allowed by the genes, but only at the cost of great effort and psychic pain, much as walking on one's knees is physically possible but is rendered quite tiring and painful by the anatomical constraints of the human body. Certain states of society are more "natural" and therefore easier and more stable. Others require a constant input of energy to maintain. Happiness is doing what comes naturally. This is the meaning of the assertion that "some behaviors can be altered experientially without emotional damage or loss in creativity. Others cannot."[64] Presumably, the price of sexual equality is eternal vigilance. To support such a concept of the genetic and physiological linkage of psychic and social states requires rather more than its simple assertion, however. Hidden behind the assertion is a complete yet unstated theory of the structure of the central nervous system for which absolutely no evidence exists. Rather than deriving its notions of psychic ease from any available knowledge of the nervous system, sociobiology has clearly inherited the idea from typological notions of normalcy and preferred natural states that were characteristic of pre-Darwinian biology.

Sometimes sociobiologists attempt to escape from the accusation that they are naive genetic determinists by explicitly stating that environment is more important than genes. The notion that genes have "given away most of their sovereignty"[65] or "hold culture on a leash"[66] simply cannot be framed in the language of genetics in a way that has any exact technical meaning.

Finally, when hard pressed, sociobiologists will sometimes say that "genes promoting flexibility in social behavior are strongly selected."[67] While that might indeed be true, it deprives sociobiology of all content. The theory must do better than simply say that human beings are adaptive machines with very complicated nervous systems.

Sociobiology offers several weak arguments for the existence of genetic control of social structures. First, the putative universality of a character is taken as evidence for its genetic control. "In hunter-gatherer societies men hunt and women stay at home. This strong bias presents in most agricultural and industrial societies and, on that ground alone, appears to have a genetic origin."[68] This argument confuses the observation with the explanation. If its circularity is not evident, one might consider the claim that, since 99 percent of Finns are Lutherans, they must have a gene for it.

A related but more serious argument is based on the supposed similarity between human social behavior and that of other primates. Evolutionary biologists distinguish between *homologous* structures, which have been inherited from common ancestors, and *analogous* structures, which may be similar in function but arise from quite different evolutionary sources. So, the wings of birds and bats are homologous, since they are formed from the forelimbs of vertebrates, while the wings of birds and insects are only analogous. If several closely related forms all have the same trait, it is reasonable to suppose they have inherited it from a recent common ancestor. However, humans have no very close living relatives. No other species is classified in the same genus (*Homo*) or family (Hominidae) or superfamily (Hominoidea), although this may simply reflect the fact that it is *Homo* who classified the animals, and the most recent common ancestor of *Homo sapiens* with the great apes was at least 2 million years ago.* Moreover, the human brain increased in volume about fourfold in that period. It is simply not possible to say that traits that *appear* to be homologous between humans and apes are really so. Behavior that is genetically stereotyped in apes may be learned early in humans. It is well known that in birds the development of the song may be genetically stereotyped in one species while in other, related forms the song must be learned. The chaffinch will sing a characteristic species song even if raised in isolation, although it does so imperfectly if it does not

*The 2-million-year estimate is a minimum based on immunological similarities between humans and great apes. The fossil evidence puts the date much farther back, at 5 million years.

hear an adult singing. The bullfinch, on the other hand, will learn to mimic an immense variety of songs and will sing the song it learns from its father, no matter what that song may be. As in much else, sociobiologists use contradictory claims to make their point. So it is said that conservative traits, similar between species, are evidence of genetic control, yet it is also asserted that labile traits are those that are most likely to differ genetically between human groups. Finally, all hope of using the evidence from similarity is compromised by admitting that we cannot be sure whether labile traits might not be homologous between humans and chimpanzees, and vice versa.[69] In fact, evidence from similarity can be used to support any argument arbitrarily.

The other evidence offered in support of genetic control of human social behavior is the assertion that a number of human traits—such as introversion-extroversion, sports activity, personal tempo, neuroticism, dominance, and schizophrenia—have been claimed to be moderately heritable. This argument is wrong in two ways. First, there simply are no adequate studies of the heritability of human personality traits. As we discussed in Chapter 5, it is critical not to confuse familial similarity with heritability. In the absence of controlled adoption studies of reasonable sample size, it is not possible to say what the causes of similarity of relatives may be. In the United States the highest correlations between parent and offspring for any social traits are for religious sect and political party. Only the most vulgar hereditarian would suggest that Episcopalianism and Republicanism are directly coded for in the genes. Nothing reveals the advocacy nature of works on sociobiology better than their cavalier treatment of the evidence on heritability of human psychosocial traits. Some quote secondary and tertiary sources of heritability estimates with no critical examination,[70] while others assure us that such human traits are indeed heritable while citing only experiments on flies, ducks, and mice.[71]

Second, heritability of a trait is evidence for genetic variation within the population, not for genetic homogeneity. As we pointed out in Chapter 5, a trait for which all individuals are genetically identical will have a heritability of zero, because heritability is a measure of the proportion of variation in a population that arises from genetic differences. Human genetics does not have any method for detecting the presence of genes controlling behavioral characters if these genes are identical in everyone. This raises the question of whether sociobiologists believe that human beings are genetically uniform for the genes of "human nature." If so, then these genes will not be detected by heritability studies. If not, then in what does genetically controlled human nature consist? If only some people have genes for aggression, both aggression and nonaggression are part of human nature.

Despite the hedges and contradictions, genetic determinism lies at the core of sociobiological theory. To make the theory work it is necessary to invoke genes with exactly the desired physiological and developmental properties to fit each case. When Owen Glendower boasted, "I can call spirits from the vasty deep," he was properly answered by Henry Percy, "Why so can I, or so can any man, / But will they come when you do call for them?" But everything we know about the development of organisms and the nature of genes tells us that there are some restrictions on the possible kinds of genetic variation that can arise within species. One is surely not at liberty to invent genes with arbitrary and complex properties for the convenience of theories. No vertebrate has ever sprouted an extra pair of limbs, and although it might be nice to have wings as well as hands and feet, the set of vertebrate genotypes does not include that possibility.*

There is a more fundamental problem for biological human nature theories. Suppose that developmental biology were to reach the point where the developmental response to environment of specific human genotypes could be specified with respect to behavior. Under those circumstances, the characteristics of an individual could be predicted, given the environment. But the environment is a social environment. What is it that determines the social environment? Somehow the characteristics of individuals are relevant, although they are not determinative. There is thus a dialectical relation between individual and society, each being a condition of the other's development and determination. The theory of this dialectical relation, in which individuals both make and are made by society, is a social theory, not a biological one. The laws of relation of individual genotype to individual phenotype cannot by themselves provide the laws of the development of society. In addition, there must be the laws that relate the collection of individual natures to

*J. B. S. Haldane once remarked that we would never become a race of angels because we lack the genetic variation for wings and for moral perfection.

the nature of the collectivity. This problem of social theory disappears in a reductionist world view, because to a reductionist, society is determined by individuals with no reciprocal path of causation.

The last element in the sociobiological argument is to reconstruct a plausible story for the origin of human social traits by natural selection. The general outline is to suppose that in the evolutionary past of the species there existed some genetic variation for a particular trait, but that the genotypes determining a particular form of behavior somehow left more offspring. As a consequence, these genotypes increased in the species and eventually came to characterize it. As an example, it is supposed that at some time in the evolutionary past some males were more genetically individualistic and less prone to accept indoctrination into group values than other males. Such nonindoctrinable males would be excluded by the group, would lose their protection in bad times, not get to share in group resources, and perhaps even be killed by their fellows. As a result, the nonindoctrinable genotypes would survive less well and leave fewer offspring, so that genetically controlled indoctrinability would become characteristic of the species. Similarly imaginative stories have been told for ethics, religion, male domination, aggression, artistic ability, etc. All one need do is predicate a genetically determined contrast in the past and then use some imagination, in a Darwinian version of Kipling's *Just So Stories*. The only trouble with Kipling was that he believed in the inheritance of acquired characteristics.

An amusing but not atypical example is a teaching exercise contrived by three prominent sociobiological anthropologists in order to teach secondary school students the elements of sociobiological reasoning.[72] They ask, "Why do children so often dislike spinach, while older people usually like it?" First the students are told how to establish the generality of this bit of human nature by asking their parents and their friends whether it is true. Then they are given the adaptive story. Spinach contains oxalic acid, which prevents the absorption of calcium. Children have growing bones and need calcium. Adults' bones are no longer growing, so the lack of calcium is not so important. Thus, any gene that had the effect of making children dislike spinach, but adults like it, would be favored. The reader should not be put off by the silliness of the case. It has all the necessary elements: (1) the appeal to everyday ethnocentric experience as evidence for universality; (2) the unstated assumption that genes may arise with any arbitrarily complicated action needed by the theory; (3) the invention of an adaptive story without any quantitative check on whether there is indeed an effect of eating spinach on reproductive rates.

A central role in the sociobiological argument about natural selection is played by Darwin's theory of sexual selection. According to this theory, males compete for females, who in turn choose from among the competitors the one whose attributes seem most likely to guarantee a large and healthy family. One is reminded of the Victorian image of the swain kneeling at the feet of his beloved, putting at her disposal all his worldly goods. The asymmetry of competition between the sexes is thought to arise from the asymmetry of their investment (note the terminology) in the production of offspring. (See also Chapter 6). Females incubate the young either internally or as eggs in a nest, and devote a good deal of their life energy to feeding and raising the young. Males are not tied down but, having contributed their infinitesimally small sperm, are free to go off and court yet other females. As a consequence, natural selection favors those females who are most careful in their choice of healthy and vigorous males to produce healthy and vigorous offspring. Males, on the other hand, are selected either to be particularly attractive to females in their coloration, song, posture, and other adornments, or else to be able to vanquish other male suitors by being more aggressive, having larger antlers, and so on.

The theory of sexual selection is a particularly flexible and powerful form of adaptive argument and has been wielded with great ingenuity by sociobiologists, in what Barash has called, with unusual candor, playing "Let's Pretend."[73] As an example of how sociobiological theory can explain anything, no matter how contradictory, by a little mental gymnastics, let us consider the paradox of feminine adornment and male drabness in the human species. The theory of sexual selection predicts that, in general, males should be the more brightly colored and highly adorned, while females should be drab, as is in fact the case among most bird species. Yet, in Western culture at least, the reverse seems true. Does this falsify the theory of sexual selection? Not at all. It is, according to Symons's *The Evolution of Human Sexuality*, just what one would expect. Females' probable reproductive success is advertised by their outward appearance (large breasts, wide hips), which women will then accentuate. Male drabness, on the other hand, demonstrates that the male is conservative and therefore likely to be a good provider economically. Moreover, males who adorn themselves are likely to be promiscuous and may abandon their families. Finally, women have been selected to be sexually attractive as a means of controlling men. "In the West, as in all human societies, copulation is usually a female service or favor."[74] (In reading sociobiology one has the constant feeling of being a voyeur, peeping into the autobiographical memoirs of its proponents.) Since "hominid females evolved in a milieu in which physical and political power was wielded by adult males,"[75] "women evolved to use their assets to their own advantage."[76]

Finally, if none of these arguments is convincing, we are reminded that Western environments are artificial, so perhaps human sexual behavior is temporarily nonadaptive, and the problem disappears.

It sometimes seems obvious that a common trait should lower rather than raise its carriers' reproductive fitness. In particular, altruistic acts that benefit others at the actor's expense should be selected against, yet altruism exists. To explain altruism, sociobiologists use stories of kin selection, part of a broader concept of extended fitness introduced by W. D. Hamilton to explain social behavior.[77] An individual's relatives have a probability of carrying the same genes as the individual, and this probability increases as the closeness of the relationship increases. Sibs have half their genes in common, first cousins only one-eighth in common. The gene for a particular trait could increase in a population if an individual carrier lowered its own reproductive fitness but at the same time increased the fitness of a relative by a large enough amount to more than compensate. So, an individual that sacrificed itself completely for three sibs would, by this indirect path, increase copies of its own genes. A variety of traits are explained as a consequence of kin selection, when direct selection seems to fail. The classic example is the explanation of homosexuality.[78] Since, it is stated, homosexuals "necessarily" leave fewer offspring than heterosexuals, the trait should disappear. It is proposed, however, that during human evolution homosexuals, not having their own families to support, devoted their energy to helping their sibs raise children, and this compensated for the homosexuals' own loss of reproductive potential and maintained their genes in the species. This story is typically superficial. First, it is by no means certain that homosexuals leave fewer offspring. While persons who are exclusively homosexual are necessarily nonreproductive, many people engage in both homosexual and heterosexual behavior. We know nothing of their reproductive rates. If one is in the business of telling unsubstantiated stories, it would be easy to claim that bisexuals are more sexually active generally. Second, there is no acceptable evidence that homosexuality has any genetic basis. Third, no evidence is actually offered that homosexuals really do (or did in the human evolutionary past) increase the reproductive rates of their sisters and brothers. And finally, the entire saga is posited on the assumption that homosexuality is the reified property of an individual, rather than an aspect of sexual expression profoundly reflective of contemporary social and cultural mores. The story has been manufactured out of whole cloth. In fact, even though a number of cases of cooperative behavior between relatives is known in various animals, in none of these has it been shown that such cooperation compensates for the loss of fitness of the cooperative actors.

The tremendous increase in the ease of adaptive storytelling that is accomplished by adding kin selection as a way of explaining individually nonadaptive traits is insufficient to deal with cases of altruism toward strangers. To take care of these cases, Trivers[79] has produced the theory of reciprocal altruism. If genes exist that induce altruistic acts toward strangers, and if these strangers remember the act and reciprocate in the future, then, provided the probabilities are right, the two altruists may gain fitness. So, if A takes a 5 percent chance of dying to save B from a 50 percent chance of death, B may do the same for A in the future, and both will benefit. Their genes for reciprocal altruism will then increase. No actual example is ever offered, so the theory remains an ingenious mental game.

The combination of direct selection, kin selection, and reciprocal altruism provides the sociobiologist with a battery of speculative possibilities that guarantees an explanation for every observation. The system is unbeatable because it is insulated from any possibility of being contradicted by fact. If one is allowed to invent genes with arbitrarily complicated effects on phenotype and then to invent adaptive stories about the unrecoverable past of human history, all phenomena, real and imaginary, can be explained. Even the most reductionist of sociobiologists sometimes become conscious of the possibility that adaptive storytelling belongs more in the realm of games than of natural science. Dawkins confesses that "there is no end to the fascinating speculation which the idea of reciprocal altruism engenders when we apply it to our own species. Tempting as it is, I am no better at such speculation than the next man, and I leave the reader to entertain himself."[80]

The central place that adaptive stories have in sociobiological explanation is in revealing contradictions with the basic claim to scientific novelty of the sociobiological method. According to sociobiology, previous theorists of the evolution of social behavior have foundered because they had too narrow a view of natural selection. Previous theory had always asked whether the possession of a trait increased or decreased the reproductive fitness of the individual possessor. This led to the paradox of the evolution of altruistic traits, which should dwindle, a paradox that was solved in older theory by postulating selection among populations. Sociobiologists quite correctly point out, however, that what matters is whether the *genes* increase in frequency in the species, so that indirect selection, as for example kin selection, can cause the increase in a trait even though its possessor is not in any sense better adapted. What is ironic is that, far from being novel, the rejection of direct adaptation as the sole motive force of evolution has been a major strain in evolutionary genetics for nearly half a century. Moreover, sociobiology completely ignores the kinds of nonadaptive explanations that are common in modern evolutionary genetics and confines itself precisely to adaptive arguments, sometimes indirect and tortured, that were characteristic of the vulgar Darwinists of the nineteenth century.

There are a number of evolutionary forces that are clearly nonadaptive and which may be correct explanations for any number of actual evolutionary events. First, there are multiple selective outcomes possible when more than a single gene influences a character. The existence of multiple adaptive states means that for a fixed regime of natural selection there are alternative paths of evolution. Which one is taken by a population depends upon chance events, so that it is not meaningful to ask for an adaptive explanation of the difference between two populations at two different outcomes of the same selective process.

For example, there is no adaptive explanation required for the existence of the two-horned rhinoceros in Africa and the one-horned rhinoceros in India. We do not have to invent an ingenious explanation of why two horns are better in the West and one in the East. Rather, they are alternative outcomes of the same general selective process. In general, nonlinear multidimensional dynamic processes have more than a single possible stable state.[81]

Second, the finite size of real populations results in random changes in gene frequency, so that, with a certain probability, genetic combinations with lower reproductive fitness, or with no differential fitness at all, will be fixed in a population.[82] If fitness differences between genotypes are small, there is a very high probability of the loss of favorable genes. This is especially true during times of restriction of population size, which is precisely when environment is likely to be changing and selective processes for new genotypes most likely to appear. Even in an infinite population, because of the nature of Mendelian genetics, a new favorable gene with a reproductive advantage s has only a probability of $2s$ of being incorporated into the population. Thus natural selection often fails to incorporate favorable genes.

Third, many changes in characters are the consequence of the multiple phenotypic effect of genes (or pleiotropy). It would be silly to argue that blood is red because redness, per se, is advantageous to the organism. Rather, the oxygen-carrying characteristics of hemoglobin are of advantage, and hemoglobin happens to be red. A special but important case of pleiotropy is allometric growth of different body parts. In cervine deer, antler size increases more than proportionately to body size as deer grow, so that larger deer have more than proportionately large antlers.[83] It is then unnecessary to give a specifically adaptive reason for the immense size of antlers of large deer.

Finally, there is an important random noise component in development and physiology. The phenotype is not given by genotype and environment alone, but is subject to the random noise processes at the molecular and cellular level. In some cases—the development of hairs in the fruit fly, for instance—variation from developmental noise may be as great as genetic and environmental variation.[84] All variation, especially in human social behavior, is not to be explained deterministically and cannot be taken as demanding specifically adaptive stories.

Sociobiological explanation, despite its claim to be the mechanical working out of the consequences of Mendelism and Darwinism, never makes use of any of these alternative modes of explanation. It would be totally foreign to sociobiology to suggest that some aspect of human behavior is simply the incidental effect of other anatomical and neurological changes or, worse, a consequence of the random fixation of genes. Sociobiologists begin with the trait and invent an origin for it which assumes that the trait itself is the efficient cause of its evolution. There is no hint in sociobiological theory that evolutionary geneticists are in serious doubt about what fraction of evolutionary change is the result of natural selection for specific characters.[85]

Given the explicit claims of sociobiology to be the extension of Darwinist and Mendelian mechanism, its contradictory devotion, in practice, to the adaptive mode of argument can only be understood as flowing from an independent ideological basis. By arguing that each aspect of the human behavioral repertoire is specifically adaptive, or at least was so in the past, sociobiology sets the stage for legitimation of things as they are. We are the products of eons of natural selection. Dare we, in our hubris, try to go against the social arrangements that nature, in its wisdom, has built into us? There is a reason why we are entrepreneurial, xenophobic, territorial. These qualities are not the consequence of blind chance, perhaps maladaptive from their very inception. This biological Panglossianism, although not a logical requirement for the biological determinist argument of inevitability, has played an important role in legitimation. More, by emphasizing that even altruism is the consequence of selection for reproductive selfishness, the general validity of individual selfishness in behaviors is supported. E. O. Wilson has identified himself with American neoconservative libertarianism,[86] which holds that society is best served by each individual acting in a self-serving manner, limited only in the case of extreme harm to others. Sociobiology is yet another attempt to put a natural scientific foundation under Adam Smith. It combines vulgar Mendelism, vulgar Darwinism, and vulgar reductionism in the service of the status quo.

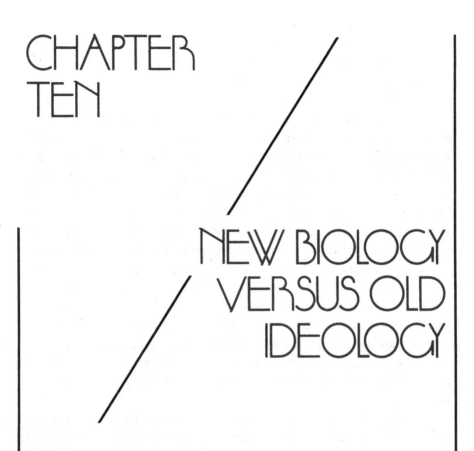

CHAPTER TEN

NEW BIOLOGY VERSUS OLD IDEOLOGY

Gene, Organism, and Society

Critics of biological determinism are like members of a fire brigade, constantly being called out in the middle of the night to put out the latest conflagration, always responding to immediate emergencies, but never with the leisure to draw up plans for a truly fireproof building. Now it is IQ and race, now criminal genes, now the biological inferiority of women, now the genetic fixity of human nature. All of these deterministic fires need to be doused with the cold water of reason before the entire intellectual neighborhood is in flames. Critics of determinism seem, then, to be doomed to constant nay-saying, while readers, audiences, and students react with impatience to the perpetual negativity. "You keep telling us about the errors and misrepresentations of determinists," they say, "but you never have any positive program for understanding human life." In the words of Lumsden and Wilson, defending their *Genes, Mind and Culture*[1] against those who accuse it of extreme determinist reductionism, critics should "fish or cut bait."[2]

We are at a severe disadvantage. Unlike the biological determinists who have simple, even simplistic, views of the bases and forms of human existence, we do not pretend to know what is a correct description of all human societies, nor can we explain all criminal behavior, wars, family organization, and property relations as manifestations of one simple mechanism. Rather, our view is that the relation between gene, environment, organism, and society is complex in a way not encompassed by simple reductionist argument. But we do not stop our analysis by simply throwing up our hands and saying that it is all too complicated for analysis. Instead, we want to propose an alternative world view. It provides a framework for an analysis of complex systems that does not murder to dissect, but that maintains the full richness of interaction that inheres in the system of relationships. Before we can begin that task of construction, however, we must revert briefly to our old negativism and again make clear what it is that we are not proposing.

A claim of biological determinists is that their critics are extreme cultural determinists. By cultural determinism they mean the view that individuals are simply mirrors of the cultural forces that have acted on them from birth. Cultural determinism can be taken to include Skinnerian behaviorism, which views individual human personality as a directly determined consequence of the sequence of sensory inputs, responses, rewards, and punishments into the developing human being from birth. Cultural determinists are also said to believe that the organism at birth is a *tabula rasa*, a blank sheet on which parents, sibs, teachers, friends, and society in general can write anything at all. So, the philosopher Midgley, herself no sociobiologist, takes it as a self-evident refutation of most antisociobiological writings that her children were patently different at birth, and so not *tabulae rasae*.[3] A corollary of extreme cultural determinism is that individuals ought to reflect accurately their family circumstances and their social class in their own behavior. We ought to be able to predict the actions of persons from their social histories. Since it is evident that we cannot make such predictions, at least in many cases, naive cultural determinism clearly is wrong. It is then asserted that we are forced back into the belief in some causal role for genes or else a mystical, nonmaterialistic belief in free will.[4] Serious supporters of extreme cultural determinism—Skinnerian behaviorists, for example—can escape this dilemma by asserting that the observations are too gross. The individual influences of our parents, teachers, and friends interact in a complicated but deterministic way to produce what appears on the surface to be exceptional behavior, but which will ultimately be analyzable into a behavioral program. Nor would such determinists claim that the newborn infant is a *tabula rasa*, since there must be some base of native abilities or properties which are then modulated by reinforcement during childhood.

The contrast between biological and cultural determinisms is a manifestation of the nature-nurture controversy that has plagued biology, psychology, and sociology since the early part of the nineteenth century. Either nature plays a determining role in producing the similarity and differences among human beings, or it does not, in which case, what is left but nurture? We reject this dichotomy. We do assert that we cannot think of any significant human social behavior that is built into our genes in such a way that it cannot be modified and shaped by social conditioning. Even biological features such as eating, sleeping, and sex are greatly modified by conscious control and social conditioning. The sexual urge in particular may be abolished, transformed, or heightened by life history events. Yet, at the same time, we deny that human beings are born *tabulae rasae*, which they evidently are not, and that individual human beings are simple mirrors of social circumstances. If that were the case, there could be no social evolution.

The materialist doctrine that men are the products of circumstances and upbringing, and that, therefore, changed men are products of other circumstances and changed upbringing, forgets that it is men that change circumstances and that the educator himself needs educating.[5]

Moreover, it is perfectly obvious that human social life is related to human biology. As we have pointed out, were human beings only six inches tall there could be no human culture at all as we understand it. Extreme cultural determinism is as absurd as its biological bedfellow. Of course, neither biological nor cultural determinists ever wish *en-*

tirely to exclude the significance of the other. Wilson, Barash, Dawkins, and others allow that if we wish (by mechanisms not biologically specified) we can transcend our genetic restrictions and create different types of (more egalitarian) societies—though we do so at our peril. Cultural determinists do not entirely deny that the biology of an infant or an aged person affects their social and cultural existence in ways that differ from those of a young adult. Both sides, however, seem to share in a type of arithmetical fallacy which argues that causes of events in the life of an organism can be partitioned out into a biological proportion and a cultural proportion, so that biology and culture together add up to 100 percent. This belief permeates not merely the exercise of attaching spurious meanings to heritability studies but also to that of diagnosing the origins and treatments for individual mental states. Depression, for example, is seen in this model as either endogeneous —caused by biological events within the individual—or exogenous—precipitated by events in the individual's external environment. Such either-or dichotomies are a logical necessity if one is bound by determinist thinking, which maintains the discrete, separable and noninterpenetrating nature of phenomena.

A second, more pluralistic, response to biological determinism is interactionism. According to this view it is neither the genes nor the environment that determines an organism but a unique interaction between them. Interactionism is the beginning of wisdom. Organisms do not inherit their traits but only their genes, the DNA molecules that are present in the fertilized egg. From the moment of fertilization until the moment of its death, the organism goes through the historical process of development. What the organism becomes at each moment depends both on the genes that it carries in its cells and on the environment in which development is occurring. Identical genotypes in different environments will have different developmental histories, just as different genotypes in the same environment will develop differently. There are no generalities that hold consistently about the ways in which different genotypes will develop differently in different environments. It all depends.

The fundamental concept for understanding the relationship between gene, environment, and organism is the norm of reaction. The norm of reaction of a genotype is the list of phenotypes that will result when the genotype develops in different alternative environments. It can be represented as a graph showing how a character of the organism changes as a function of its environmental experience. Each different genotype is characterized by its own norm of reaction, and there is no simple relationship among these norms. For example, one genotype may grow better than a second at a low temperature, but more poorly at a high temperature. A well-documented example is the relative performance of hybrid corn varieties. All corn hybrids improve their yield as the amount of nitrogen, water, and sunlight is increased, but some respond more than others. A curious consequence of these different reaction norms in response to environmental amelioration is that modern corn hybrids are superior to those of fifty years ago when tested at high planting densities in somewhat poorer environments, while the older hybrids are superior at low planting densities and in enriched conditions. Plant breeding has then not selected for "better" hybrids but for hybrids that do better than the older varieties under stress conditions but more poorly than old varieties when both are given superior growing conditions. Thus genotype and environment interact in a way that makes the organism unpredictable from a knowledge of some average of effects of genotype or environment taken separately. We are in no doubt that, were the processes of development sufficiently well understood, and given a sufficient amount of detailed information about the genotype of an organism, we could predict the phenotype in any given environment. But we do not have such knowledge, or anywhere near it, so that for the forseeable future only empirical observation can reveal what norms of reaction look like.

No one has ever measured the norm of reaction for any human genotype, because to do so would require the replication of that genotype in many fertilized eggs and then placing the developing infants, all genetically identical, into a variety of deliberately chosen environments. Nevertheless, judging from what is known of reaction norms from experimental plants and animals, it is overwhelmingly likely that human norms of reaction are constant over some ranges of environment and change their relative positions over others. As an example, consider body temperature. At room temperature, when fully clothed, all healthy human beings have virtually identical body temperatures, 37° Centigrade. If they are stripped, however, and sent out into freezing weather, thin people will suffer a loss in body temperature much more quickly than fat ones. In contrast, if they are required to do heavy work in the sun, the fat will experience a dangerous rise in temperature before the thin. Body conformation is known to have some heritability, but whether a heritable difference in body conformation makes a difference to heat regulation, and the direction of that difference depends on the environment.

At first sight, interactionism, with its recognition of the unique interaction between genes and environment in determining the organism, would seem to be the correct alternative to biological or cultural determinism. It has the seductive appeal of a middle way that does not sacrifice a basic commitment to cause-and-effect determinism, nor even to reductionism, but restates the empirical problem as that of uncovering the mechanism of environmental influence on the developing genotype. How does nitrogen affect the rate of synthesis of certain plant proteins whose cellular control is under the influence of specific genes? Indeed, in the example of human body temperature, we already have the physiological model that explains the differential response of fat and thin people to temperature stress. Yet interactionism, while a step in the right direction, is flawed as a mode of explanation of human social life. It carries with it two basic assumptions that it has in common with more vulgar determinisms and that prevent its solving the problem of society. First, it supposes the alienation of organism and environment, drawing a clean line between them and supposing that environment makes organism, while forgetting that organism makes environment. Second, it accepts the ontological priority of the individual over the collectivity and therefore of the epistemological sufficiency of the explanation of individual development for the explanation of social organization. Interactionism implies that if only we could know the norms of reaction of all living human genotypes and the environments in which they find themselves we would understand society. But in fact we would not.

The Organism as Respondent

There have been two powerful metaphors that have characterized both biological and social theory. The first, older metaphor is that of unfolding, or unrolling, that is etymologically hidden in the English word "development"* but more transparent in the Spanish *desarollo*. Organisms, societies, cultures are seen as containing all that they ever are to be immanent in their earliest form, and requiring only an initial triggering to set them off on their preset path of developmentally unfolding. Often this unfolding is described in terms of stages that succeed each other in fixed order, whether they be the golden, silver, bronze, and iron ages of civilization seen by the Greeks; the oral, anal, and genital progression of Freud; or the sensorimotor, preconceptual, operational, and formal stages of Piaget.

With such models comes the notion of arrested development, so that individuals may be "fixed" in, say, their anal stage and never progress beyond it. For instance, the theory of neoteny supposes that some species reach adulthood sooner in the course of development than others and so resemble the juvenile stages of the other species. Humans are much more similar in morphology to fetal than to adult apes. We are, so to speak, apes born too soon. Theories of unfolding give supremacy to internal factors of development, reserving for the environment only the role of triggering the process or of blocking its further progress at one stage or another. It is thus itself a biological determinist model.

The newer metaphor, introduced for the first time in the nineteenth century, is a unique intellectual contribution of Darwin. It is the metaphor of trial and error, of challenge and response, of problem and solution. In this model, organisms, societies, and species confront problems set for them by external nature, independent of their own existence, and they respond by trying various solutions until one is found that fits. The archetype is the variational model of Darwinian evolution. The external world poses problems of survival and reproduction. Species adapt by throwing up random variants, the "trials," some of which succeed reproductively, spreading through the species, providing the species with an adaptive response to the external challenge. The same metaphor appears in theories of cultural evolution. Cultures vary one from another in their ways of confronting the environment. Some, like us, made the right guess, while others, like the Fuegians, were less fit culturally and died out. Or else, particular cultural forms or ideas—the "memes" of Dawkins[6]—have superior

---

*And in the word "evolution" as well, which originally meant an unfolding of the immanent.

reproductive power. Christianity vanquishes heathenism because it is more appealing to the mind and better meets the demands of life.

Trial and error has also become the metaphor for a number of theories of psychological development and learning, the evolutionary epistemologies of Popper, Lorenz, Campbell, and Piaget.[7] Children in their development (or indeed sciences as well, according to Popper) meet problems set by the external world. They make conjectural solutions to these problems, which are tested against nature, refuted by experience, and replaced by other conjectures. Finally, a system of knowledge that most closely approximates a true perception of nature is built up by trial and error. There are multiple developmental pathways possible for the individual organism. Internal factors, such as the genes, only generate the conjectures. The organism develops psychically by constantly referring these conjectures to the environment, which determines which will be acceptable. This, then, is an interactionist model.

The feature that is common to both the unfolding and the trial-and-error metaphors is the asymmetric relation between organism and environment. The organism is alienated from the environment. There is an external reality, the environment, with laws of its own formation and evolution, to which the organism adapts and molds itself, or dies if it fails. The organism is the subject and the environment is the object of knowledge. This view of organism and environment pervades psychology, developmental biology, evolutionary theory, and ecology. Changes in organisms both within their lifetimes and across generations are understood as occurring against a background of an environment that has its own autonomous laws of change and that interacts with organisms to direct their change. Yet, despite the near universality of this view of organism and environment, it is simply wrong, and every biologist knows it.

Interpenetration of Organism and Environment

The problem with trying to describe an autonomous environment is that there is an infinity of ways in which the bits and pieces of the world can be put together to make environments. We must make a clear distinction between an unstructured external world of physical forces, and the environment (literally, the surroundings) of an organism, which is defined by the organism itself. In the absence of actual organisms, how can we know which combination of factors is an environment and which is not? In fact, it is the organisms themselves that define their own environment. A practical example of the importance of the organism in defining its environment is in the design of the Mars lander that was to detect life on that planet. The lander carried an artificial environment consisting of a nutrient soup and a set of instruments that would detect the production of carbon dioxide when that soup was metabolized by Martian life. But such an instrument defines life as something that can be sucked up into the soup and will break it down to produce carbon dioxide. The marvelous irony, totally unforeseen, was that the soup was broken down and gas evolved from it, but in a pattern over time that was unlike anything seen on Earth. After a year of soul-searching debate, biologists finally decided that it was not life at all but a previously unknown form of inorganic reaction taking place in clay particles sucked into the machine. The designers of the lander had constructed a Martian environment based on their knowledge of terrestrial organisms and so, in effect, had accepted those organisms' definition of the environment.

Organisms do not simply adapt to previously existing, autonomous environments; they create, destroy, modify, and internally transform aspects of the external world by their own life activities to make this environment. Just as there is no organism without an environment, so there is no environment without an organism.* Neither organism nor environment is a closed system; each is open to the other. There is a

*It is interesting that in his latest book, *The Extended Phenotype,*[8] Dawkins has endeavored to come to grips with the environment. True to his reductionist principles, he is forced to handle the fact that the organism acts on its environment by defining what we here call the "active environment" as an aspect of the organism's phenotype. Thus the dam a beaver constructs becomes part of the beaver's phenotype; the lake is "determined" by the beavers' genes. Even organisms become part of one anothers' phenotype. Thus viruses make us sneeze so as to increase the chance of infecting another host; air travel becomes a phenotypic manipulation by disease-producing organisms so as to increase their own spread. The whole argument explodes into caricature; everything vanishes into the maw of the DNA serpent, which pulls itself slowly inside out and reveals at last to the startled world—precisely the organism and its interpenetrations that Dawkins has tried to magic away!

variety of ways in which the organism is the determinant of its own milieu.

First, organisms construct their environments out of bits and pieces of the world. The dead straw in a garden is part of the active environment of a phoebe because the phoebe gathers that straw to make its nest. The stones in the garden are not part of the phoebe's active environment, although they are in direct physical proximity to the straw, but they are part of the active environment of a thrush, which uses them to break snails on. Neither the straw nor the stones are part of the active environment of a woodpecker that lives in the dead beech at the foot of which both stones and straw lie. Which pieces of the world are relevant and how these relevant bits are related to each other in the life of an organism change as the organism itself develops, either in its lifetime or in evolutionary time. All living plants and animals are covered in a thin layer of warm air created by their metabolism. A small parasite, say a flea, that lives on the skin of an animal is submerged in that warm boundary layer which constitutes part of its environment. Should the flea grow larger, however, it will emerge from that air mantle into the cold stratosphere a few millimeters away from the animal's skin. It will have put itself into a new environment. While it is a commonplace that human beings can reconstruct their environment at will, it is not always appreciated that environmental construction is a universal feature of all life.

Second, organisms transform their environments. Not only human beings but all living beings both destroy and create the resources for their own continued life. As plants grow, their roots alter the soil chemically and physically. The growth of white pines creates an environment that makes it impossible for a new generation of pine seedlings to grow up, so hardwoods replace them. Animals consume the available food and foul the land and water with their excreta. But some plants fix nitrogen, providing their own resources; people farm; and beavers build dams to create their own habitat. Indeed, a significant part of the natural history of New England is a consequence of the actions of beavers raising and lowering the water table.

Third, organisms transduce the physical nature of environmental inputs. Changes in external temperature are felt by one's body organs not as heat but as change in the concentration of certain hormones and sugar in one's blood. When one sees and hears a rattlesnake, the photon and molecular energy that excites one's eyes and ears is sensed by one's internal organs as a change in adrenaline concentration. Presumably the effect of the same sight and sound on another snake rather than on a human would be quite different.

Fourth, organisms alter the statistical pattern of environmental variation. Fluctuations in food supply are damped out by storage devices. The potato tuber is a damping device for the plant that human beings have captured for their own purposes. But small differences can also be magnified, as when our central nervous systems pick out a signal from background noise because our attention is called to it. Organisms integrate fluctuations to record only the total, as for example plants that flower only after a sufficient number of accumulated degree-days above a critical temperature have been experienced.

The point of this survey of the nature of interactions between organisms and their environments is that all organisms—but especially human beings—are not simply the results but are also the causes of their own environments. Development, and certainly human psychic development, must be regarded as a codevelopment of the organism and its environment, for mental states have an effect on the external world through human conscious action. While it may be true that at some instant the environment poses a problem or challenge to the organism, in the process of response to that challenge the organism alters the terms of its relation to the outer world and recreates the relevant aspects of that world. The relation between organism and environment is not simply one of interaction of internal and external factors, but of a dialectical development of organism and milieu in response to each other.

Critics of the generality of the dialectical relation between organism and environment sometimes claim that important aspects of nature are not accessible to change. After all, we did not pass the law of gravity; we are stuck with it as a universal fact of nature. Yet gravitation is precisely an example of how the nature of the organism determines the relevance of a "universal fact of nature." Larger numbers of aquatic microorganisms and soil bacteria live "outside" of gravity because their tiny size makes their weight irrelevant as far as gravity is concerned. However, these organisms are severely buffeted by a "universal physical force", Brownian motion, in the surrounding water molecules that we, at our relatively enormous size, are totally unaware of and unaffected by. There is no universal physical fact of nature whose effect on, or even relevance to an organism is not in part a consequence of the nature of the organism itself.

What is true of organisms in general is all the more accentuated in human psychic development. At every instant the developing mind, which is a consequence of the sequence of past experiences and of internal biological conditions, is engaged in a recreation of the world with which it interacts. There is a mental world, the world of perceptions, to which the mind reacts, which at the same time is a world created by the mind. It is obvious to all of us that our behavior is in reaction to our own interpretations of reality, whatever that reality may be. We perceive others to be hostile, friendly, intelligent, stupid, generous, or mean, and can do so almost independently of their objective behavior or their own self-perception.

Further, our behavior in response to that self-created mental world recreates the objective world that surrounds us. If we perceive others constantly as hostile to us and behave toward them as if they were hostile, they indeed become so, and the perception becomes reality. As a child develops, its psychic environment comes into being partly as a consequence of its own behavior. And all successful scientists know that as they become more and more successful and are given greater and greater recognition, any foolish or shallow statements they make are more and more likely to be given credence and even invested with a depth they do not have. The result is an increase in the scientists' self-esteem and in their public reputation. That is not to deny that the psychic environment has a certain autonomy as well. As a character of Saul Bellow observed, "Just because I am paranoid doesn't mean that people don't persecute me." Nevertheless, a remarkably large contribution to our psychic and social environments is created by and in response to our own actions. Thus, any theory of psychic development must include not only a specification of how a given biological individual develops psychically in a given sequence of environments, but how the developing individual in turn interpenetrates with the objective and subjective worlds to recreate its own environments.

The alienation of organism and environment in biological and social theory, despite its obvious falsity, is a double consequence of ideological developments that we have previously discussed. Subject and object are separated as part of the reductionist metaphysic, while all interactions in the world are seen as asymmetrical, between identifiable subject and object. It is this feedback that distinguishes interactionism from our view of the interpenetration of organism and environment. Interactionism takes the autonomous genotype and an autonomous physical world as its starting point and then describes the organism that will develop from this combination of genotype and environment. But nowhere is it recognized that in the process that external world is reorganized and redefined in its relevant aspects by the developing organism.

The hierarchical nature of human social organization makes the subject-object dichotomy seem only natural when we contemplate the physical world. But that alienation is also of direct political relevance. The alienated organism must accommodate itself to the facts of life: "That's life, so you'd better learn to live with it." Accommodation as a political goal is hypostasized as a concrete, necessary relation between organisms and their environments, quite outside their control. Thus, psychic maturation is seen as learning to replace wishes about the world with acceptance of its actual nature. In the words of Piaget:

Adolescent egocentricity is manifested by a belief in the omnipotence of reflection, as though the world should submit itself to idealistic schemes rather than to systems of reality. . . .

Equilibrium is attained when the adolescent understands that the proper function of reflection is not to contradict but to predict and interpret experience.[9]

To which we can only juxtapose Marx's famous eleventh thesis on Feuerbach: "The philosophers have only interpreted the world in various ways; the point, however, is to change it."

Levels of Organization and Explanation

A second failure of interactionism, like that of cultural and biological determinism, is that it is unable to come to grips with the fact that the material universe is organized into structures that are capable of analysis at many different levels. A living organism—a human, say—is an assemblage of subatomic particles, an assemblage of atoms, an assemblage of molecules, an assemblage of tissues and organs. But it is not first a set of atoms, then molecules, then cells; it is all of these at the same time. This is what is meant by saying that the atoms, etc., are not ontologically prior to the larger wholes that they compose.

Conventional scientific languages are quite successful when they are confined to descriptions and theories entirely within levels. It is relatively easy to describe the properties of atoms in the language of physics, of molecules in the language of chemistry, of cells in the language of biology. What is not so easy is to provide the translation rules for moving from one language to another. This is because as one moves up a level the properties of each larger whole are given not merely by the units of which it is composed but of the organizing relations between them. To state the molecular composition of a cell does not even begin to define or predict the properties of the cell unless the spatiotemporal distribution of those molecules, and the intramolecular forces that are generated between them, can also be specified. But these organizing relationships mean that properties of matter relevant at one level are just inapplicable at other levels. Genes cannot be selfish or angry or spiteful or homosexual, as these are attributes of wholes much more complex than genes: human organisms. Similarly, of course it makes no sense to talk of human organisms showing base pairing or Van der Waal's forces, which are attributes of the molecules and atoms of which humans are composed. Yet this confusion over levels and the properties appropriate to them is one that determinism constantly gets involved in.

Consider the types of explanation that it is possible to offer of a relatively straightforward biological event, the contraction of a muscle in a frog's leg. One can offer as a cause for the twitch of the muscle an explanation couched completely in the language of physiology: The muscle twitched because an appropriate set of impulses passed down the motor nerve innervating it, which signaled the instruction to contract. Here the present phenomenon is caused by an immediate prior event: First the nerve fires, then the muscle twitches—and one can go on to explain that the nerve fired as a result of some earlier appropriate set of input to its motor neurons, derived from the frog's brain and/or its sensory inputs. So we have a sequential series of events that follow one another in time and are linked in a transitive and irreversible way.

First, event A occurs; as a result, event B; as a result, event C; and so on. This is a straightforward causal chain, all the individual components of which are described in the same language and within a single level of analysis. The sequence is that of Figure 10.1. The single-headed arrows emphasize that one could not run the sequence backwards, so to say; the muscle twitch cannot cause the events in the motor nerve to occur.

But this is not the only way of explaining the muscle twitch. One can also consider the activity of the whole organism and then state that the muscle twitched because the frog was jumping to escape a predator. Here the explanation of the activity of part of a complex system is given in terms of the integrated functioning of the system as a whole. Systems approaches appear to give a meaning to the activity that cannot be derived or understood from a single-level approach, by defining it in terms of the goals of the organism.

Such holistic explanations are the source of much confusion; indeed, such "general systems theorists" as Paul Weiss, Ludwig von Bertlanffy, or Arthur Koestler have ascribed almost mystical significance to them.[10] In his effort to avoid either a reductionist or a dualist trap, the neurophysiologist Roger Sperry, for example, claims that they represent a form of "downward causation" by which the properties of the system—the organism—constrain or determine the behavior of the parts.[11] The system thus becomes more important than the parts of which it is composed. If an experimenter severed the motor nerves to the frog's leg muscle or paralyzed the muscle with a chemical poison, the frog would still endeavor to escape from its predator—and possibly succeed—by employing a different set of muscles or a different escape strategy.

To the goal-directed organism there are multiple paths to a given end. Some even argue that to bother about the exact mechanisms involved is irrelevant to achieving a main understanding of what is happening. In the examples often provided, one does not need to know the mechanism of the internal combustion engine or how a silicon chip works to drive a car or use a pocket calculator. What is clear, though, is that any account of the structure of a muscle cell in the leg of a frog that ignores the fact that it is part of a system for moving the limb relative to the rest of the body is just inadequate. Simply to catalogue all the parts of which a car is composed and their interactions would

FIGURE 10.1.

$\longrightarrow$ sensory inputs $\longrightarrow$ brain events $\longrightarrow$ motor output $\longrightarrow$ muscle twitch

TIME

$\longrightarrow$

not tell you anything about the function of the car, what it was like to drive, or its role in a transportation system.

Holistic explanations bear a sort of mirror image relationship to reductionism. Let us return to the frog muscle. It is itself composed of individual muscle fibers. These themselves are largely composed of fibrous proteins. In particular, there are two protein molecules, actin and myosin, arranged within the muscle fibrils in characteristic arrays. When the muscle fibrils contract, the actin and myosin chains slide between each other—a series of molecular conformational changes that involve the expenditure of energy. For reductionism, the muscle twitch is *caused* by the proteins sliding between one another, and reductionism would seek to go on to explain the protein movements in terms of the properties of the molecular and atomic constituents of those proteins.

But just as there are not two successive phenomena—*first* the frog jumping, *then* the muscle twitch—there is not *first* the sliding of the protein molecules and *then* the twitch. The sliding molecules *constitute* the twitch, but at the biochemical rather than the physiological level of analysis. While within-level causal explanations describe a temporal sequence of events, reductionist and holistic accounts alike are not causal in this sense at all; they are different descriptions of a unitary phenomenon. A full and coherent explanation of the phenomenon requires all three types of description, but without giving primacy to any one.

Actually, for completeness other types of description are also required: The properties of the muscle cannot be understood except in the context of the development of the individual frog from the egg to the adult, which defines the relationship of the parts of the frog as an organism. And the part played by the twitching muscle in the survival of the frog and the propagation of its kind cannot be understood except

FIGURE 10.2 / Types of causal explanation in biology

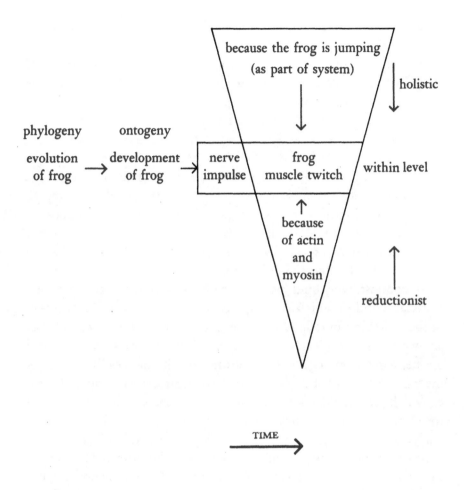

in reference to the evolution of frogs in general (or phylogeny).[12] The relationship of these sets of descriptions within the explanation of the frog muscle twitch is given in Figure 10.2.

It is this integration which is misunderstood by interactionism, which confuses the epistemological plurality of levels of explanation with the ontological assumption that there are really many different and incompatible types of cause in the real world. Such an assumption either leads to an empty mysticism or generates paradox. Consider

Sperry's argument, referred to above, for "downward causation." If all he is saying is that, within a complex whole, the degrees of freedom available to the component parts are differently determined than if the parts were isolated monads, he is obviously right. But it is clear he means more than just this. He means that there are two types of radically incommensurable causes determining the behavior of the parts of any system. Some run "up" as when the interdigitation of muscle proteins "cause" the muscle to contract. Others run "down" as when the instruction "jump" causes the contraction. Presumably causes pass one another as they cross levels, like commuters going respectively up and down on parallel escalators. The image conveys the paradox which is always present at the heart of such dualism, for how can ontologically different types of causation each produce an identical set of results? Perhaps it is for this reason that when faced with real methodological challenge Sperry-type holism collapses so easily into hard-nosed reductionism.

By contrast, we would insist on the unitary ontological nature of a material world in which it is impossible to partition out the "causes" of the twitching muscle of the frog into x percent social (or holistic) and y percent biological (or reductionist). The biological and the social are neither separable, nor antithetical, nor alternatives, but complementary. All causes of the behavior of organisms, in the temporal sense to which we should restrict the term *cause*, are simultaneously both social and biological, as they are all amenable to analysis at many levels. All human phenomena are simultaneously social and biological, just as they are simultaneously chemical and physical. Holistic and reductionist accounts of phenomena are not "causes" of those phenomena but merely "descriptions" of them at particular levels, in particular scientific languages. The language to be used at any time is contingent on the purposes of the description; the muscle physiologist is interested in a different aspect of the question of the frog-muscle twitch from the ecologist or evolutionary biologist or biochemist; their difference of purpose should define the language of description to be used.

Minds and Brains

Nowhere is the confusion between levels of analysis and levels of reality more apparent than in the discussion of the relationship of minds to brains. Brains, for reductionists, are determinate biological objects whose properties produce the behaviors we observe and the states of thought or intention we infer from that behavior. Minds, according to the dominant position of Western philosophy, the so-called central-state materialism, may simply be reduced to brains. Mind events (thoughts, emotions, and so forth) are caused by brain events, or can be regarded simply as rather unsatisfactory and unscientific ways of talking about those events.

Such a position is, or ought to be, completely in accord with the principles of sociobiology offered by Wilson and Dawkins. However, to adopt it would involve them in the dilemma of first arguing the innateness of much human behavior that, being liberal men, they clearly find unattractive (spite, indoctrination, etc.) and then to become entangled in liberal ethical concerns about responsibility for criminal acts, if these, like all other acts, are biologically determined. To avoid this problem, Wilson and Dawkins invoke a free will that enables us to go against the dictates of our genes if we so wish. Thus Wilson allows that despite the genetic instructions that demand male domination, we *can* create a less sexist society—at the cost of some loss of efficiency[13]—and goes on to speculate on the evolution of culture.[14] Dawkins offers independently evolving cultural units, or memes.[15]

This is essentially a return to unabashed Cartesianism, a dualistic *deus ex machina*. It is, incidentally, the position to which a number of neuroscientists, whose lifelong research techniques on the brain have been unremittingly reductionist, have also returned. Neurophysiologist Sir John Eccles argues that, residing in the left hemisphere of a hard-wired and determinate brain there is to be found a region—as yet unapproached by his electrodes—called the liaison brain, which is in direct communication with a disembodied mind which can exert its will over the brain's machinery.[16] Neurosurgeon Wilder Penfield, after many years of electrically stimulating the brains of epileptic patients and evoking movements, sensations, and memories, claimed a similar seat of the mind.[17]

Such dualism is an attempted escape for those who cannot see how to get themselves out of the corner into which they have deterministically painted themselves. In the case of the neuroscientists it comes from the fallacy of spurious localization—that there must be a site in the brain in which consciousness, like a homunculus, resides. To this argument we respond that the property of being a mind—of "minding"—must be seen as the activity of the brain as a whole; the product of the interactions of all of its cellular processes with the external world. To think otherwise is analogous to making the mistake of believing that we can see because in the visual cortex of our brains there is located a camera taking pictures of the images on the retina, together with a miniature observer to scan and interpret the pictures. On the contrary, the total of the activity of the cells of the visual system of the brain *is* the act of seeing and interpreting what we see.

In the case of sociobiologists the dualism arises from that other reductionist error—the inappropriate partitioning of causes. "If I lift my arm above my head, either that is free will or it is biologically determined." But "free will" is the name given to a set of mental processes. Such processes, like those of the lifting of arm, can also be described in physiological language. The confusion over free will arises entirely from the misplacing of levels of causation and levels of analysis. Our actions are no more to be partitioned between free will and determinism than are our bodies or brains to be partitioned between nurture and nature. To say that we have simultaneously minds and brains, and are simultaneously social and biological, is to transcend these false dichotomies and to point the way toward an integrated understanding of the relationship between our conscious and biological selves.

From Individual to Society

Thus, interactionism first asymmetrically separates subject and object, and second, confuses the levels of analysis of the relations of subject and object. The third failure of interactionism as a response to determinism is that it confuses the collection of individual norms of reaction with social organization.

Suppose that we knew all the genotypes present in a human population, and suppose further that we knew the norms of reaction of each one so that we could specify the psychic development of every individual in any given family and social environment. How are we to take the collection of these predicted individual psychic developments and convert them into a prediction of society? To do so, we would need more than the complete biological theory of norms of action. We would need, in addition, a purely social theory that converts the collection of individuals into an organized society.

Both biological determinism and interactionism implicitly possess such a theory. It is the assumption—an extension of the analytic confusion that we have referred to above—that social properties are a direct compositional consequence of the collection of individual properties. We go to war because we are a collection of aggressive people, so war is to be prevented, if at all, only by making each of us peaceable. We have organized religion because each of us has a religious impulse. We have rich and poor because some of us have ability and others do not. Sometimes a more sophisticated version of this compositional theory is advanced. Perhaps only a small critical mass of people with a given characteristic is needed before the entire society takes on a property of these leaders. Only a few influential religious or aggressive persons will be sufficient, under such a theory, to convert the society as a whole.[18] Yet it is not difficult to show that such compositional theories cannot be correct.

In the first place, there are many properties of social organization that are allocational and that cannot be altered by changing the composition of the population. Thus it is not the case that the proportion of persons in different professions, trades, skills, services, and labor processes is a consequence of the proportion of different available skills. The number of physicians is determined by the available places in medical schools, a number not set by any study of the number of able people but by the economics of the profession. As we pointed out earlier, it would be absurd to suppose that if only bankers had children, everyone would be a banker. It is important to realize that these allocational properties would apply even if people with different qualities are differentially allocated. Even though only the tallest people can get on a professional basketball team, an increase in the average height of the population will not increase the number of teams, only the average height of the players. While biological determinists sometimes claim that the allocation into social positions and skills is limited by natural availability,[19] the high unemployment among graduate engineers, Ph.D.s in English, philosophy, history, and the like clearly shows this to be nonsense.

In the second place, historical changes in social structure have taken place with such rapidity that no alteration in proportions of different genotypes in the population can possibly explain them. The rise of the people of Arabia and the Maghreb in the hundred years following the Hegira from a poor, backward, pastoral, and local merchant society to the great civilization of Mediterranean Islam, preeminent in poetry, mathematics, science, and political power, can hardly be explained by a change in gene frequencies.*

In the third place, individual human constraints do not appear at the level of social organization. One of the chief claims of sociobiology is that society is constrained by individual properties that are translated as prohibitions on society. Yet the most striking feature of social life is that it so often is the negation of individual limitations. Indeed, that negation is the force that keeps societies together. People can do in concert what they cannot do separately. Nor is this property simply the result of the summation of individual forces, as when ten people can lift a weight that one person alone cannot. On the contrary, totally new properties arise from social interaction. None of us can fly by flapping our arms either singly or in a crowd. Yet we do fly as a result of technology, airplanes, pilots, airlines, ground crew, all *de novo* products of social activity, qualitatively different from our individual acts. Moreover, it is not society that flies, but individuals. The memories of individuals are limited, and if all the historians in the world were set to the task, they could not learn by rote even a tiny fraction of the factual material (the census figures, for instance) they use in their profession. Yet they can recall these facts, as individuals, by going to the library and reading books, a qualitatively new product of social activity. Once again, individuals acquire new properties from society.

At the same time, society is obviously made up of individuals. Society is not, in a metaphor that has persisted in various forms through

*Not that biological determinists haven't had a try. For an absurd, tailor-made model that attempts this, see Lumsden and Wilson's *Genes, Mind and Culture.*

many centuries, itself an organism. It is not a Platonic form that has an independent existence above and outside of individual people. It is their creation. It is, as Marx said, "men that change circumstances." While Newtonian mechanics would have come into existence even if Isaac Newton had died in his crib, it was, in fact, a product of individual thought. Society does not think; only individuals think. Thus, the relation between individual and society, like the relation between organism and environment, is a dialectical one. It is not only that society is the environment of the individual and therefore perturbs and is perturbed by the individual. Society is also hierarchically related to individuals. As a collection of individual lives, it possesses some structural properties, just as all collections have properties that are not properties of the individuals that make them up, while at the same time lacking certain properties of the individuals. Only an individual can think, but only a society can have a class structure. At the same time, what makes the relation between society and the individual dialectical is that individuals acquire from the society produced by them individual properties, like flying, that they did not possess in isolation. It is not just that wholes are more than the sum of their parts; it is that parts become qualitatively new by being parts of the whole.

Determination and Freedom

Dialectical determination is still determination, and so, like the biological determinists, we must confront the problem of freedom. If all effects have causes (at least above the level of quantum mechanics), then what can we mean by freedom in a material, causal world? If any choice one makes is a consequence of one's mental state at the moment of choice, and if mental states are part of a natural chain of causation from antecedent conditions, is one really free?

For biological as well as for behaviorist determinists, the answer is no. While we may have illusions of freedom, they claim that in fact our choices are programmed by our genes or our infant training. We are, in Dawkins's phrase, "lumbering robots" containing genes that "control us, body and mind."[20] Even the illusion of freedom has been programmed into us by evolution because illusions are adaptive. "Men would rather believe than know,"[21] the author of *Sociobiology* asserts authoritatively.

For most moral philosophers the answer has been yes, but the problem of reconciling freedom with a belief in causation has been a troublesome one. For Kant the solution, if it can be called that, was to accept an unreconciled dual nature of human existence. As material beings, we are totally caused and, so, totally determined. But as social and moral beings, we are free to make choices and must bear the responsibility for our acts. Hume's solution was to move the problem onto a more political and practical terrain. We are free, he held, when we act according to our desires and wishes. The prisoner is not free because, although he wishes to be at large, he is physically restrained by outside forces. The madman is not free because he is constrained by a pathological compulsion. Whether desires and wishes are themselves a consequence of an antecedent chain of natural cause is not relevant to this view of freedom. Yet Hume's solution is somehow unsatisfying. Freedom ought to have the property of transitivity. If we act according to our wishes and desires, but those wishes and desires were in some way programmed by our genes and past experiences, then our actions were so programmed at second hand.

At one level we cannot but ally ourselves with Hume. Any theory of freedom that cannot distinguish between one's freedom to leave one's house and go downtown and a prisoner's inability to do so, or between a rich person's freedom to take a Caribbean vacation and a poor one's necessity to stay at home and shiver, is both absurd and a political obfuscation. Whatever Big Brother may have said, slavery is not freedom. At a deeper level, however, we must try to understand freedom of choice as a consequence of, rather than as a contradiction to, causation.

If we look at physical systems, we see that randomness and determination, far from being in contradiction, arise one from the other as levels of organization are crossed. The decay of radioactive nuclei is truly causeless and random in the sense that there is no difference in state between a nucleus that will or will not decay up until the actual instant of radioactive emission. Yet the most exquisitely accurate clocks, precise to a millionth of a second, are those that use the number of random radioactive emissions per second as their counters. Conversely, the movement of a microscopic particle in a gas is random in any practical sense in which we can mean that word, yet it is the consequence of a large number of deterministic collisions and is completely specified by those events.

It is usually said that the randomness of nuclear disintegration and the randomness of molecular movement are distinct, the first being true ontological randomness, the second having only the appearance of randomness because of our limited knowledge of the antecedent conditions. If only we could see all the bombarding molecules and calculate their pathways, we could predict the path of our particle exactly. But this claim of epistemic randomness obscures a vital physical difference between a bouncing molecule and, say, a train on a railroad track. The movement of the molecule is a consequence of the conjunction of a vast multitude of causal chains, each independent of the other, all of which intersect to produce the history of the particle. The consequence is that the path of the particle is itself only infinitesimally correlated with any one of the intersecting chains of causation. While the path is totally determined by the ensemble of causes, it is essentially independent of any one of them. Not so the train, whose path is entirely constrained by the tracks. Yet the train is moving at random with respect to the motions of people in the towns through which it passes, although there is an infinitesimal gravitational attraction between them. That is, when we speak of randomness, we must specify randomness with respect to what phenomenon. What we mean by randomness is, in fact, independence of one action from another.

What characterizes human development and actions is that they are the consequence of an immense array of interacting and intersecting causes. Our actions are not at random or independent with respect to the totality of those causes as an intersecting system, for we are material beings in a causal world. But to the extent that they are free, our actions are independent of any one or even a small subset of those multiple paths of causation: that is the precise meaning of freedom in a causal world. When, on the contrary, our actions are predominantly constrained by a single cause, like the train on the track, the prisoner in his cell, the poor person in her poverty, we are no longer free. For biological determinists we are unfree because our lives are strongly constrained by a relatively small number of internal causes, the genes for specific behaviors or for predisposition to these behaviors. But this misses the essence of the difference between human biology and that of other organisms. Our brains, hands, and tongues have made us independent of many single major features of the external world. Our biology has made us into creatures who are constantly re-creating our own psychic and material environments, and whose individual lives are the outcomes of an extraordinary multiplicity of intersecting causal pathways. Thus, it is our biology that makes us free.

NOTES

CHAPTER ONE / THE NEW RIGHT AND THE OLD DETERMINISM

1. For a discussion of New Right ideology see, e.g.: P. Green, *The Pursuit of Inequality* (New York: Pantheon Books, 1981). P. Steinfels, *The Neo-Conservatives* (New York: Simon & Schuster, 1979) for the U.S. For the U.K. and Thatcherism: M. Barker, *The New Racism* (London: Junction Books, 1981), and the series of articles in *Marxism Today:* M. Jacques, October 1979, pp. 6–14; S. Hall, February 1980, pp. 26–28; I. Gough, July 1980, pp. 7–12.

2. K. Marx and F. Engels, *The German Ideology* (1846), chap. 1, pt. 3, art. 30 (New York: International Publishers, 1974).

3. R. Nisbet, quoted in Jacques, *op. cit.*

4. A. Ryan, "The Nature of Human Nature in Hobbes and Rousseau," in *The Limits of Human Nature,* ed. J. Benthall (London: Allen Lane, 1973), pp. 3–20.

5. For a robust defense of reductionism in biology and psychology, see, e.g.: M. Bunge, *The Mind Body Problem* (Oxford: Pergamon, 1981); M. Boden, *Purposive Explanation in Psychology* (Cambridge, Mass.: Harvard Univ. Press, 1972); E. Wilson, *The Mental as Physical* (London: Routledge & Kegan Paul, 1979); F. Crick, *Life Itself* (London: Macdonald, 1982); J. Monod, *Chance and Necessity* (London: Cape, 1972); and S. Luria, *Life: The Unfinished Experiment* (London: Souvenir Press, 1976).

6. *The Guardian* (London), 14 July 1981.

7. Claimed in two articles by National Front theoretician R. Verrall in *The New Nation*, nos. 1 and 2 (summer and autumn 1980).

8. R. Dawkins, defending himself and sociobiology against the charge of giving succor to racist and fascist ideologies in *Nature* 289 (1981): 528.

9. R. Dawkins, *The Selfish Gene* (Oxford: Oxford Univ. Press, 1976), p. 126.

10. On this vexed topic, see, for example, H. Rose and S. Rose, eds., *The Political Economy of Science* (London: Macmillan, 1976), and *The Radicalisation of Science* (London: Macmillan, 1976).

11. Science for the People, *Biology as a Social Weapon* (Minneapolis, Minn.: Burgess, 1977).

12. E. O. Wilson, *Sociobiology: The New Synthesis* (Cambridge, Mass.: Harvard Univ. Press, 1975).

13. For instance, "antipsychiatrists" like T. Szasz in *The Manufacture of Madness* (London: Routledge & Kegan Paul, 1971); D. Ingleby, *Critical Psychiatry: The Politics of Mental Health* (Harmondsworth, Middlesex, England: Penguin, 1981); M. Foucault, *Madness and Civilization* (London: Tavistock, 1971); and his followers such as J. Donzelot, *The Policing of Families: Welfare versus the State* (London: Hutchinson, 1980).

14. It is interesting that even an archetypal biological determinist like Dawkins has sooner or later to come to grips with the environment. His latest book, *The Extended Phenotype* (London: Freeman, 1981), is a long struggle to reduce even an organism's environment to a product of its "selfish genes."

CHAPTER TWO / THE POLITICS OF BIOLOGICAL DETERMINISM

1. A. R. Jensen, "How Much Can We Boost IQ and Scholastic Achievement?" *Harvard Educational Review* 39 (1969): 1–123.

2. R. J. Herrnstein, *IQ in the Meritocracy* (Boston: Little, Brown, 1971).

3. H. J. Eysenck, *Race, Intelligence and Education* (London: Temple Smith, 1971), and *The Inequality of Man* (London: Temple Smith, 1973). These books were followed by such National Front pamphlets, drawing explicitly upon them, as *How to Combat Red Teachers* (London, 1979).

4. E. O. Wilson, "Human Decency Is Animal," *New York Times Magazine*, 12 October 1975, pp. 38–50.

5. V. H. Mark and F. R. Ervin, *Violence and the Brain* (New York: Harper & Row 1970).

6. See T. Powledge, "Can Genetic Screening Prevent Occupational Disease?" *New Scientist*, 2 September 1976, p. 486; D. J. Kilian, P. J. Picciano, and C. B. Jacobson in "Industrial Monitoring, a Cytogenetic Approach," *Annals of the New York Academy of Sciences* 269 (1975); J. Beckwith, "Recombinant DNA: Does the Fault Lie Within Our Genes?" *Science for the People* 9 (1977): 14–17.

7. H. Rose "Up Against the Welfare State: The Claimant Unions," in *Socialist Register*, ed. R. Miliband and J. Saville (London: Merlin Press, 1973), pp. 179–204.

8. W. Ryan, *Blaming the Victim* (New York: Pantheon Books, 1971).

9. E. Zola, Preface to *La Fortune des Rougon*, Librairie Internationale, A. Lacrois (Paris: Verboeckhoven, 1871).

10. See, for example, H. F. Garrett, *General Psychology* (New York: American Book, 1955).

11. C. Lombroso, *L'homme criminal* (Paris: Alcan, 1887).

12. P. A. Jacobs, M. Brunton, M. M. Melville, R. P. Brittan and W. F. McClamont, "Aggressive Behaviour, Mental Subnormality and the XYY Male," *Nature* 208 (1970): 1351–52.
For a survey of the literature on XYY and aggression see R. Pyeritz, H. Schrier, C. Madansky, L. Miller, and J. Beckwith, "The XYY Male: The Making of a Myth," in *Biology as a Social Weapon* (Minneapolis: Burgess, 1977). For a discussion of this progression see S. Chorover *From Genesis to Genocide* (Cambridge, Mass.: MIT Press, 1979).

13. For the history of the relations between genetics, eugenics, and statistics see D. A. MacKenzie, *Statistics in Britain, 1865–1930* (Edinburgh: Edinburgh University Press: 1981).

14. Quoted by R. Hofstadter, *Social Darwinism in American Thought* (New York: Braziller, 1959).

15. The history of the IQ testing movement in the United States is given in, for example: L. Kamin, *The Science and Politics of IQ* (Potomac, Md: Erlbaum, 1974); A. Chase, *The Legacy of Malthus* (Urbana: University of Illinois Press, 1980); D. P. Pickens, *Eugenics and the Progressives* (Nashville: Vanderbilt Univ. Press, 1968); J. M. Blum, *Pseudoscience and Mental Ability* (New York: Monthly Review Press, 1978); D. L. Eckberg, *Intelligence and Race* (New York: Praeger, 1979); and K. M. Ludmerer, *Genetics and American Society* (Baltimore: Johns Hopkins Univ. Press, 1972). For the United Kingdom, see N. Stepan, *The Idea of Race in Science* (London: Macmillan, 1982); B. Evans and B. Waites, *IQ and Mental Testing* (London: Macmillan, 1981); and also see the famous UNESCO *Statement on Race*, whose main author was Ashley Montagu (Montagu, 1950).

16. See R. Verrall, *New Nation*, Summer 1980. For France, J. Brunn, *La Nouvelle Droite* (Paris: Oswald, 1978); "J. P. Hebert" (pseud.), *Race et intelligence* (Paris: Copernic, 1977).

17. L. Agassiz, "The Diversity of Origin of the Human Races," *Christian Examiner* 49 (1850): 110–45.

18. B. Davis, "Social Determinism and Behavioural Genetics," *Science* 189 (1975): 1049.

19. L. Agassiz quoted in W. R. Stanton, *The Leopard's Spots: Scientific Attitudes Towards Race in America* (Chicago: Univ. of Chicago Press, 1960), p. 106.

20. Wilson, *Sociobiology*, p. 575. For other attempts to derive ethics from biology, for example, see V. R. Potter, *Bioethics* (Englewood Cliffs, N.J.: Prentice-Hall, 1972); and G. E. Pugh, *The Biological Origin of Human Values* (New York: Basic Books, 1977).

21. K. Lorenz, "Durch Domestikation verursachte Stölunchen arteigenen verhaltens," *Zeit für Angewandte Psychologie und Characterkunde* 59 (1940): 2–81.

22. F. Galton, *Inquiries into Human Faculty and Its Development*, 2nd ed. (New York: Dutton, 1883).

23. For the debate about the status of scientific theories see, e.g.: I. Lakatos and A. Musgrave, eds., *Criticism and the Growth of Knowledge* (Cambridge: Cambridge Univ. Press, 1970), L. Laudan, *Progress and Its Problems* (Berkeley: Univ. of California Press, 1977); R. Bhaskar, *A Realist Theory of Knowledge* (Hassocks, Sussex, England: Harvester, 1978).

24. For discussion of the social context of science and scientific knowledge see, for example, H. Rose and S. Rose, *The Political Economy of Science* (London: Macmillan, 1976). Also H. Rose and S. Rose, "Radical Science and its Enemies," in *The Socialist Register*, ed. R. Miliband and J. Saville (1979): pp. 317–35.

CHAPTER THREE / BOURGEOIS IDEOLOGY AND THE ORIGIN OF DETERMINISM

1. C. B. Macpherson, *The Political Theory of Possessive Individualism* (New York: Oxford University Press, 1962).

2. This correspondence was first pointed out, in an essay that was to change the shape of the subsequent historiography of science, by Boris Hessen in *Science at the Crossroads*, ed. N. Bukharin et al. (Moscow: Kniga, 1931).

3. For example, J. R. Ravetz, *Scientific Knowledge and Its Social Problems* (London: Allen Lane, 1972). Also H. Rose and S. Rose, *Science and Society* (Harmondsworth, Middlesex, England: Penguin, 1969).

4. A. Sohn-Rethel, *Mental and Manual Labour* (London: Macmillan, 1978).

5. C. Dickens, *Hard Times* (Penguin Edition, 1969), pp. 48, 126.

6. On the theme of the domination of nature see W. Leiss, *The Domination of Nature* (Boston: Beacon, 1974). Also, A. Schmidt, *The Concept of Nature in Marx* (London: New Left Books, 1973).

7. A. Pope, *Moral Essays*, Epistle 1 to Lord Cobham.

8. H. Driesch, *The History and Theory of Vitalism* (London: Macmillan, 1914); also see J. S. Fruton, *Molecules and Life* (New York: John Wiley, 1972).

9. R. Virchow, *The Mechanistic Concept of Life* (1850), trans. in *Disease, Life and Man*, ed. J. K. Lelland (Stanford, Calif.: Stanford Univ. Press, 1958). Also see J. Loeb, *The Mechanistic Concept of Life*, reprinted with an introduction by D. Fleming (Cambridge, Mass.: Harvard Univ. Press, 1964).

10. K. Marx, "Theses on Feuerbach," (1845) in K. Marx and F. Engels, *Selected Works* vol. 1 (Moscow: Progress Publishers, 1969)

11. Stated by biochemist W. L. Byrne at a conference on "Learning disability," Kansas City, 1979.

12. F. Jacob, *The Logic of Living Systems* (London: Allen Lane, 1974).

13. See, for example, R. M. Young, *Mind, Brain and Adaptation in the Nineteenth Century* (New York: Oxford Univ. Press, 1970).

14. S. J. Gould, *The Mismeasure of Man* (New York: Norton, 1981).

15. B. L. Priestly and J. Lorber, "Ventricular Size and Intelligence in Achondroplasia," *Zeitschrift für Kinderchirurgie* 34 (1981): 320–26.

16. C. Lombroso, quoted in S. Chorover, *From Genesis to Genocide* (Cambridge, Mass.: MIT Press, 1979), pp. 179–80.

17. Chorover, *From Genesis to Genocide*, p. 180.

18. A. Christie, *The Secret Adversary* (New York: Dodd, Mead, 1922), p. 49.

19. For example, A. T. Scull, *Museums of Madness: The Social Organisation of Insanity in 19th Century England* (London: Allen Lane, 1979).

20. C. Darwin, *The Expression of the Emotions in Man and Animals* (London: John Murray, 1872).

21. F. Galton, *Hereditary Genius* (London: Macmillan, 1969).

22. For example: Gould, *Mismeasure of Man*. Also see A. Chase, *The Legacy of*

Malthus (Urbana: Univ. of Illinois Press, 1980); Chorover, *From Genesis to Genocide;* and B. Evans and B. Waites, *IQ and Mental Testing* (London: Macmillan, 1981).

23. For Crick on the "Central Dogma" see F. H. C. Crick, *Symposium of the Society for Experimental Biology* 12 (1957): 138–63; *Perspectives in Biology and Medicine* 17 (1973): 67–70; and *Nature* 227 (1970): 561–63.

24. J. Monod, quoted in H. Judson, *The Eighth Day of Creation* (London: Cape, 1979), p. 212.

25. H. Rose and S. Rose, "The Myth of the Neutrality of Science," in *The Social Impact of Modern Biology,* ed. W. Fuller. (London: Routledge & Kegan Paul, 1971), pp. 283–94.

26. For example, J. Hirschleifer, "Economics from a Biological Viewpoint," *Journal of Law and Economics* 20, 1 (1977): 1–52.

27. Monod, quoted in Hudson, *Eighth Day of Creation,* p. 212.

CHAPTER FOUR / THE LEGITIMATION OF INEQUALITY

1. M. Luther, *On Marriage* (1530).

2. See C. Jencks, *Inequality* (New York: Basic Books, 1972), chap. 7; see also P. Townsend, *Poverty* (Harmondsworth, Middlesex, England: Penguin, 1980).

3. Table given on p. 277 in G. M. Trevelyan, *English Social History* (New York: Longmans, Green, 1942).

4. See P. Deane and W. A. Cole, *British Economic Growth, 1688–1959* (Cambridge: Cambridge Univ. Press, 1969).

5. U.S. Bureau of the Census, *Historical Statistics of the United States: Colonial Times to 1970* (Washington, D.C.: Department of Commerce, 1975).

6. L. Doyal, *The Political Economy of Health* (London: Pluto, 1979); *The Black Report: Inequalities in Health* (DHSS London, 1980), pub. and ed. P. Townsend and N. Davidson, (Harmondsworth, Middlesex, England: Penguin, 1982).

7. L. F. Ward, *Pure Sociology* (London: Macmillan, 1903).

8. M. Young, *The Rise of the Meritocracy* (Harmondsworth, Middlesex, England: Penguin, 1961).

9. A. R. Jensen, "How Much Can We Boost IQ and Scholastic Achievement?" *Harvard Educational Review* 39 (1969): 15.

10. R. Herrnstein, *IQ and the Meritocracy,* (Boston: Little, Brown, 1973), p. 221.

11. L. F. Ward, *Pure Sociology.*

12. P. Blau and O. D. Duncan, *The American Occupational Structure* (New York: John Wiley, 1967).

13. For example, H. J. Muller, *Out of the Night* (New York: Vanguard Press, 1935).

14. T. Dobzhansky, *Genetic Diversity and Human Equality* (New York: Basic Books, 1973).

15. E. O. Wilson, *Sociobiology: The New Synthesis* (Cambridge, Mass.: Harvard Univ. Press, 1975), p. 554.

16. Ibid., p. 575.

17. Ibid.

18. For example, E. Mandel's discussion of science in *Late Capitalism* (London: Verso, New Left Books, 1978); or, for the orthodox Soviet position, M. Millionschikov in *The Scientific and Technological Revolution: Social Effects and Prospects* (Moscow:

Progress Publishers, 1972). In a curious way this determinist position becomes reflected in the writings of some of the most libertarian of the 1970s radical science movement. See for example, R. M. Young, "Science *is* Social Relations," *Radical Science Journal* 9 (1977): 61–131; also The RSJ Collective, "Science, Technology, Medicine and the Socialist Movement," *Radical Science Journal* 11 (1981): 1–70. For the critique by H. Rose and S. Rose, "Radical Science and Its Enemies," *Socialist Register,* ed. R. Miliband and J. Saville (London: Merlin, 1979), pp. 317–34.

19. G. Lukacs, *History and Class Consciousness* (London: Merlin Press, 1971).

20. A. Heller, *The Theory of Need in Marx* (London: Allison & Busby, 1977).

21. Mao Tse-tung, "On Practice," *Selected Works* (Peking: Foreign Language Press, 1962), p. 375.

22. For instance, B. Barnes and S. Shapin, *Natural Order* (London: Sage, 1979).

23. For a critique of this position, see P. Sedgwick, *Psychopolitics* (London: Pluto, 1982).

24. R. Rosenthal and L. Jacobson, *Pygmalion in the Classroom* (New York: Holt, Rinehart & Winston, 1968).

25. B. F. Skinner, *Beyond Freedom and Dignity* (London: Cape, 1972).

CHAPTER FIVE / IQ: THE RANK ORDERING OF THE WORLD

1. S. Bowles and V. Nelson, "The Inheritance of IQ and the Intergenerational Transmission of Economic Inequality," *Review of Economics and Statistics* 54, no. 1 (1974).

2. M. Schiff, M. Duyme, A. Dumaret, and S. Tomkiewicz, " 'How Much *Could* We Boost Scholastic Achievement and IQ Scores?' Direct Answer from a French Adoption Study," *Cognition* 12 (1982): 165–96.

3. A. Binet, *Les Idées modernes sur les enfants* (Paris: Flammarion, 1913), pp. 140–41.

4. L. M. Terman, "Feeble-minded children in the Public Schools of California," *School and Society* 5 (1917) 165.

5. L. M. Terman, *The Measurement of Intelligence* (Boston: Houghton Mifflin, 1916), pp. 91–92.

6. H. H. Goddard, *Human Efficiency and Levels of Intelligence* (Princeton, N.J.: Princeton Univ. Press, 1920), pp. 99–103.

7. C. Burt, "Experimental Tests of General Intelligence," *British Journal of Psychology* 3, (1909): 94–177.

8. C. Burt, *Mental and Scholastic Tests,* 2nd ed. (London: Staples, 1947); and *The Backward Child,* 5th ed. (London: Univ. of London Press, 1961).

9. L. Kamin, *The Science and Politics of IQ* (Potomac, Md.: Erlbaum, 1974); K. Ludmerer, *Genetics and American Society* (Baltimore: Johns Hopkins Univ. Press, 1972); M. Haller, *Eugenics: Hereditarian Attitudes in American Thought* (New Brunswick, N.J.: Rutgers Univ. Press, 1963); C. Karier, *The Making of the American Educational State* (Urbana.: Univ. of Illinois Press, 1973), and N. Stepan, *The Idea of Race in Science* (London: Macmillan, 1982).

10. E. G. Boring, "Intelligence as the Tests Test It," *New Republic* 34 (1923): 35–36.

11. S. Bowles and V. Nelson, "The Inheritance of IQ and the Intergenerational Reproduction of Economic Inequality," *Review of Economics and Statistics* 56 (1974): 39–51.

12. E.L. Thorndike, *Educational Psychology* (New York: Columbia Univ. Teachers College, 1903) p. 140.

13. A. R. Jensen, "Sir Cyril Burt" (obituary), *Psychometrika* 37 (1972): 115–17.

14. H. J. Eysenck, *The Inequality of Man* (London: Temple Smith, 1973).

15. C. Burt, "Ability and Income," *British Journal of Educational Psychology* 13 (1943): 83–98.

16. C. Burt, "The Evidence for the Concept of Intelligence," *British Journal of Educational Psychology* 25 (1955): 167–68.

17. H. J. Eysenck, "H. J. Eysenck in rebuttal," *Change* 6, no. 2 (1974).

18. L. Kamin, "Heredity, Intelligence, Politics and Psychology," unpublished presidential address to meeting of the Eastern Psychological Association (1972).

19. Kamin, *Science and Politics of IQ.*

20. A. R. Jensen, "Kinship correlations reported by Sir Cyril Burt," *Behavior Genetics* 4 (1974): 24–25.

21. A. R. Jensen, "How Much Can We Boost IQ and Scholastic Achievement?" *Howard Educational Review* 39 (1969): 1–123.

22. O. Gillie, *Sunday Times* (London), 24 October 1976.

23. A. R. Jensen, "Heredity and Intelligence: Sir Cyril Burt's Findings," letters to the *Times* (London), 9 December 1976, p. 11.

24. H. J. Eysenck, "The Case of Sir Cyril Burt," *Encounter* 48 (1977): 19–24.

25. H. J. Eysenck, "Sir Cyril Burt and the Inheritance of the IQ," *New Zealand Psychologist* (1978).

26. L. S. Hearnshaw, *Cyril Burt: Psychologist* (London: Hodder & Stoughton, 1979).

27. N. J. Mackintosh, Book review of *Cyril Burt: Psychologist* by J. S. Hearnshaw, *British Journal of Psychology* 71 (1980): 174–75.

28. A balance sheet on Cyril Burt, *Supplement to the Bulletin of the British Psychological Society* 33 (1980): i.

29. J. Shields, *Monozygotic Twins Brought up Apart and Brought up Together* (London: Oxford Univ. Press, 1962).

30. H.H. Newman, F.N. Freeman, and K.J. Holzinger, *Twins: A Study of Heredity and Environment* (Chicago: Univ. of Chicago Press, 1973).

31. N. Juel-Nielsen, "Individual and Environment: A Psychiatric and Psychological Investigation of Monozygous Twins Raised Apart," *Acta Psychiatrica et Neurologica Scandanavica*, Supplement 183 (1965).

32. Kamin, *Science and Politics of IQ.*

33. B. S. Burks, "The Relative Influence of Nature and Nurture upon Mental Development: A Comparative Study of Foster Parent–Foster Child Resemblance and True Parent–True Child Resemblance," *Yearbook of the National Society for the Study of Education* 27 (1928): 219–316.

34. A. M. Leahy, "Nature-nurture and Intelligence," *Genetic Psychology Monographs* 17 (1935): 235–308.

35. Kamin, *Science and Politics of IQ.*

36. S. Scarr and R. A. Weinberg, "Attitudes, Interests, and IQ," *Human Nature* 1 (1978): 29–36.

37. J. M. Horn, J. L. Loehlin, and L. Willerman, "Intellectual Resemblance Among Adoptive and Biological Relatives: The Texas Adoption Project," *Behavior Genetics* 9 (1979): 177–207.

38. R. T. Smith, "A Comparison of Socio-environmental Factors in Monozygotic

and Dizygotic Twins: Testing an Assumption," in *Methods and Goals in Human Behavior Genetics*, ed. S. G. Vandenberg (New York: Academic Press, 1965).

39. M. Skodak and H. M. Skeels, "A Final Follow-up Study of One Hundred Adopted Children," *Journal of Genetic Psychology* 75 (1949): 83–125.

40. B. Tizard, "IQ and Race," *Nature* 247 (1974): 316

41. Ibid.

42. S. Scarr-Salapatek and R. A. Weinberg, "IQ Test Performance of Black Children Adopted by White Families," *American Psychologist* 31 (1976): 726–39.

43. J. Loehlin, G. Lindzey, and J. Spuhler, *Race Differences in Intelligence* (San Francisco: Freeman, 1975).

44. Schiff et al., "How Much *Could* We Boost Scholastic Achievement" (pp. 165–96).

CHAPTER SIX / THE DETERMINED PATRIARCHY

1. We would like to acknowledge our particular debt, in writing this chapter, to the feminist scholarship on which we have drawn extensively, and in particular to the critical comments on earlier drafts made by Lynda Birke, Ruth Hubbard, and Hilary Rose.

2. Z. R. Eisenstein, ed., *Capitalist Patriarchy and the Case for Socialist Feminism* (New York: Monthly Review Press, 1979); C. Delphy, *The Main Enemy: A Materialist Analysis of Women's Oppression*, WRRC Publication no. 3 (London, 1977); M. Barrett and M. McIntosh, "The Family Wage," in *The Changing Experience of Women*, ed. E. Whitelegg et al. (Oxford: Martin Robertson, 1982); H. Hartmann, *The Unhappy Marriage of Marxism and Feminism* (London: Pluto, 1981); and A. Oakley, *Sex, Gender and Society* (New York: Harper & Row, 1972).

3. Quoted by K. Paige in "Women Learn to Sing the Blues," *Psychology Today*, September 1973; According to the *Alloa* [Scotland] *Advertiser*, at the time of the Falklands/Malvinas War in 1982, Tam Dalyell, M.P., claimed that Margaret Thatcher "was not fully capable of making vital decisions like that between war and peace simply because she was a woman and like every woman was affected by the menstrual cycle."

4. *Wall Street Journal*, 20 July 1981.

5. For example, see correspondence in the *Morning Star* (London), especially letters by M. McIntosh (24 November 1982) and B. MacDermott (27 November 1982).

6. H. Land, "The Myth of the Male Breadwinner," *New Society*, 9 October 1975; H. Rose and S. Rose, "Moving Right Out of Welfare–and the Way Back," *Critical Social Policy* 2, no. 1 (1982): 7–18.

7. Quoted in *The Sun* (London), 18 February 1981.

8. J. Morgall, "Typing Our Way to Freedom: Is it True That New Office Technology Can Liberate Women? in *Changing Experience of Women*, pp. 136–46.

9. S. Witelson, quoted in *Psychology Today*, November 1978, p. 51.

10. J. Money and A. A. Ehrhardt, *Man and Woman, Boy and Girl* (Baltimore: Johns Hopkins Univ. Press, 1972). Their list of criteria also includes expenditure of energy in outdoor play and games, fantasies of materialism and romance, and childhood sexual play.

11. J. Herman, *Father-Daughter Incest* (Cambridge, Mass.: Harvard Univ. Press, 1981);

L. Armstrong, "Kiss Daddy Goodnight," in *Speakout on Incest* (New York: Hawthorn, 1978).

12. P. L. van den Berghe, "Human Inbreeding Avoidance: Culture in Nature," *Behavioral and Brain Sciences* 6 (1983): 125–68; Also see P. P. G. Bateson, "Rules for Changing the Rules," in *Evolution From Molecules to Men*, ed. D. S. Bendall, (Cambridge: Cambridge Univ. Press, 1983).

13. K. F. Dyer, "The Trend of the Male and Female Performance Differential in Athletics, Swimming and Cycling, 1958–1976," *Journal of Biosocial Science* 9 (1977): 325–39; also see K. F. Dyer, *Challenging the Men: Women in Sport* (St. Lucia, Australia: Univ. of Queensland Press, 1982).

14. R. Hubbard, "Have Only Men Evolved?" in *Women Look at Biology Looking at Women*, ed. R. Hubbard, M. S. Henifin, and B. Fried (Cambridge, Mass: Schenkman, 1979), pp. 7–36; R. Hubbard and M. Lowe, Introduction to R. Hubbard and M. Lowe, eds., *Genes and Gender* II (New York: Gordian Press, 1979): pp. 9–34; L. Birke, "Cleaving the Mind: Speculations on Conceptual Dichotomies," in *Against Biological Determinism*, ed. S. Rose (London: Allison & Busby, 1982): pp. 60–78; and L. Rogers, "The Ideology of Medicine," in *Against Biological Determinism*, pp. 79–93.

15. H. Fairweather, "Sex Differences in Cognition," *Cognition* 4 (1976): 31–280.

16. E. E. Maccoby and C. N. Jacklin, *The Psychology of Sex Differences* (Stanford, Calif.: Stanford Univ. Press, 1974).

17. Witelson, quoted in *Psychology Today*, November 1973, pp. 48–59.

18. C. P. Benbow and J. C. Stanley, *Science* 210 (1980): 1262–64.

19. The story of this exclusion has often been told. See, for example, C. St. John-Brooks, "Are Girls Really Good at Maths?" *New Society*, 5 March 1981, pp. 411–12; A. Kelly, ed., *The Missing Half: Girls and Science Education* (Manchester: Manchester Univ. Press, 1979); N. Weisstein, "Adventures of a Woman in Science," in *Women Look at Biology Looking at Women*, pp. 187–206; M. Couture-Cherki, "Women in Physics," in *The Radicalization of Science*, ed. H. Rose and S. Rose (London: Macmillan, 1976), pp. 65–75.

20. See, for example, E. Fee, "Science and the Woman Problem: Historical Perspectives," in *Sex Differences: Social and Biological Perspectives*, ed. M. S. Teitelbaum (New York: Anchor Doubleday, 1976): pp. 173–221; J. Sayers, *Biological Politics: Feminist and Anti-Feminist Perspectives* (London: Tavistock, 1982); M. R. Walsh, "The Quirls of a Woman's Brain" in *Women Look at Biology Looking at Women*, pp. 103–26; S. S. Mosdale, "Science Corrupted: Victorian Biologists Consider the Woman Question," *Journal of the History of Biology* 11 (1978): 1–55; S. A. Shields, "Functionalism, Darwinism, and the Psychology of Women: A Study in Social Myth," *American Psychologist*, July 1975, pp. 739–54.

21. For example, C. Hutt, *Males and Females* (Harmondsworth, Middlesex, England: Penguin, 1972).

22. Fairweather, "Sex differences in Cognition."

23. Ibid.

24. L. McKie and M. O'Brien, eds., *The Father Figure* (London: Tavistock, 1982).

25. S. Rose, *The Conscious Brain* (Harmondsworth, Middlesex, England: Penguin, 1976).

26. E. Fee, "Nineteenth-Century Craniology: The Study of the Female Skull," *Bulletin of the History of Medicine* 53 (1979): 415–33.

27. Mosdale, "Science Corrupted."

28. Fee, "Nineteenth-Century Craniology;" also see D. A. MacKenzie, *Statistics in Britain, 1865–1930* (Edinburgh: Edinburgh Univ. Press, 1981).

29. Darwin, *Descent of Man*, p. 569, quoted by Mosdale, "Science Corrupted."

30. F. Pruner, in *Transactions of the Ethnological Society* 4 (1866): 13–33; quoted by Fee, "Nineteenth-Century Craniology."

31. A. R. Jensen, "A Theoretical Note on Sex Linkage and Race Differences in Spatial Visualization Ability." *Behavior Genetics* 8 (1978): 213–17.

32. N. Geschwind and P. Behan. "Left Handedness: Association with Immune Diseases, Migraine and Developmental Learning Disorder," *Proceedings of the National Academy of Sciences* 79 (1982): 5097–5100.

33. F. Nottebohm and A. V. Arnold, "Sexual Dimorphism in Vocal Control Areas of the Songbird Brain," *Science* 194 (1976): 211–13.

34. P. D. Maclean. "The Triune Brain, Emotion and Scientific Bias," in *The Neurosciences: Second Study Program*, ed. F. O. Schmitt (New York: M.I.T. Press, 1970): pp. 336–49.

35. For example, A. Koestler, *The Ghost in the Machine* (London: Hutchinson, 1967).

36. J. Jaynes, *The Origin of Consciousness in the Breakdown of the Bicameral Mind* (Boston: Houghton Mifflin, 1976); R. F. Ornstein, *Psychology of Consciousness* (New York: Harcourt Brace, 1977).

37. Witelson, quoted in *Psychology Today*, November 1978, p. 51.

38. Gina quoted by S. L. Star, "The Politics of Right and Left: Sex Differences in Hemispheric Brain Asymmetry," in *Women Look at Biology Looking at Women*, pp. 61–76.

39. S. Goldberg. *The Inevitability of Patriarchy*, 2nd ed. (New York: Morrow, 1974).

40. See Science for the People, ed., *Biology as a Social Weapon* (Minneapolis: Burgess, 1977).

41. Quoted by A. M. Briscoe in E. Tobach and B. Rosoff, eds., *Genes and Gender* (New York: Gordian Press, 1979) vol. 1, p. 41.

42. L. I. A. Birke, "Is Homosexuality Hormonally Determined?" *Journal of Homosexuality* 6 (1981): 35–49.

43. P. C. B. Mackinnon, "Male Sexual Differentiation of the Brain," *Trends in Neurosciences*, November 1978; K. D. Dohler, "Is Female Sexual Differentiation Hormone Mediated?" *Trends in Neurosciences*, November 1978.

44. E. Pizzey and J. Shapiro, *Prone to Violence* (London: Hamlyn, 1982).

45. M. Cerullo, J. Stacey, and W. Breines, "Alice Rossi's sociobiology and Antifeminist Backlash," *Feminist Studies* 4, no. 1 (February 1978); N. Chodorow, *The Reproduction of Mothering: Psychoanalysis and the Sociology of Gender* (Berkeley: Univ. of California Press, 1979).

46. L. Tiger and R. Fox, *The Imperial Animal* (London: Secker & Warburg, 1977); L. Tiger, *Men in Groups* (London: Secker & Warburg, 1969).

47. F. Engels, *The Origin of the Family, Private Property and the State* (New York: International Publishers, 1972).

48. G. Bleaney, *Triumph of the Nomads: A History of the Aborigines* (Melbourne: Overlook Press, 1982); N. M. Tanner, *On Becoming Human* (Cambridge, England: Cambridge Univ. Press, 1981).

49. N. M. Tanner, *On Becoming Human*.

50. This metaphor is used by E. O. Wilson in *On Human Nature* to epitomize his view of the relationship between genes for social behavior and manifest social relations.

51. T. R. Halliday. "The Libidinous Newt: An Analysis of Variations in the Sexual Behaviour of the Male Smooth Newt, *Triturus vulgaris,*" *Animal Behavior* 24 (1976): 398–414.

52. M. K. McClintock and N. T. Adler, "The Role of the Female during Copulation in Wild and Domestic Norway Rats (rattus Norvegicus)," *Behaviour* 68 (1978): 67–96.

53. S. Zuckerman, *The Social Life of Apes* (London: Kegan Paul, 1932); C. Russell and W. M. S. Russell, *Violence, Monkeys and Man* (London: Macmillan, 1968).

54. L. Liebowitz, *Females, Males, Families: A Biosocial Approach* (North Scituate, Mass.: Duxbury Press, 1978).

55. S. Firestone, *The Dialectic of Sex;* see H. Rose and J. Hanmer, "Women's Liberation: Reproduction and the Technological Fix," in *The Political Economy of Science,* ed. H. Rose and S. Rose (London: Macmillan, 1974), pp. 142–60.

56. See, for example, S. B. Hrdy, *The Woman That Never Evolved* (Cambridge, Mass.: Harvard Univ. Press, 1981); and E. Morgan, *The Descent of Woman* (New York: Stein & Day, 1972).

57. See, for example, J. Mitchell, *Sexual Politics* (London: Abacus, 1971).

58. H. Rose, "Making Science Feminist," in *The Changing Experience of Women,* pp. 352–72.

59. See references in note 19 above and also R. Arditti, "Women in Science: Women Drink Water While Men Drink Wine," *Science for the People* 8 (1976): 24; E. F. Keller "Feminism and Science," *Signs* 7 (1982): 589–602; A. Y. Leevin and L. Duchan, "Women in Academia," *Science* 173 (1971): 892–95; L. Curran, "Science Education: Did She Drop Out or Was She Pushed?" in *Alice Through the Microscope,* ed. Brighton Women in Science Group (London: Virago, 1980), pp. 22–41; R. Wallsgrove, "The Masculine Face of Science," in *Alice Through the Microscope,* pp. 228–40.

60. H. Rose, "Hand, Heart and Brain: Towards a Feminist Epistemology of the Natural Sciences," *Signs* (Fall 1983).

61. For a discussion of this emphasis on the domination of nature even in Marxist and radical thought, see, for example, A. Schmidt, *The Concept of Nature in Marx* (London: New Left Books, 1971); W. Leiss, *The Domination of Nature* (New York: Braziller, 1972).

62. See, for example, C. Merchant, *The Death of Nature: Women, Ecology and the Scientific Revolution* (London: Wildwood House, 1980); Boston Women's Health Book Collective, *Our Bodies, Ourselves* (New York: Simon & Schuster, 1976).

CHAPTER SEVEN / ADJUSTING SOCIETY BY ADJUSTING THE MIND

1. S. Block and P. Reddaway, *Russia's Political Hospitals: Abuse of Psychiatry in the Soviet Union* (London: Gollancz, 1977).

2. Z. A. Medvedev and R. A. Medvedev, *A Question of Madness* (London: Macmillan, 1971).

3. World Psychiatric Association, Declaration of Hawaii, *British Medical Journal* 2/6096 (1977): 1204–5.

4. J. K. Wing, "Social and Familial Factors in the Causation and Treatment of Schizophrenia," in *Biochemistry and Mental Disorder,* ed. L. L. Iversen and S. Rose (London: Biochemical Society, 1973).

5. L. Gostin, "Racial Minorities and the Mental Health Act," *Mind Out,* May 1981; *The Guardian* (London), 23 March 1981.

6. P. Bean, *Compulsory Admissions to Mental Hospitals* (London: John Wiley, 1980).

7. *New Statesman,* 3 June 1980.

8. V. H. Mark, W. H. Sweet, and F. R. Ervin, "Role of Brain Disease in Riots and Urban Violence," *Journal of the American Medical Association* 201 (1967): 895.

9. V. H. Mark and F. R. Ervin. *Violence and the Brain* (New York: Harper & Row, 1970). The quotation is from p. 7.

10. E. M. Opton, correspondence circulated at Winter Conference on Brain Research, Vail, Colorado, 1973; developed in A. W. Schefflin and E. M. Opton, *The Mind Manipulators* (London: Paddington Press, 1978); quoted in S. Rose, *The Conscious Brain* (Harmondsworth, Middlesex, England: Penguin, 1976).

11. Cited by S. Chavkin, *The Mind Stealers: Psychosurgery and Mind Control* (Boston: Houghton Mifflin, 1978).

12. " 'Rioters may be taking to the streets because of the high level of lead in their bodies,' a professor claimed yesterday," "This England," *New Statesman,* 24 July 1981. Also see papers by O. David, "The Relationship Between Lead and Hyperactivity," and H. C. Needleman, "Studies of the Neurobehavioural Costs of Low-Level Lead Exposure," presented at the Conference on Low-Level Lead Exposure and Its Effects on Human Beings (CLEAR), London 1982.

13. J. M. R. Delgado, *Physical Control of the Mind: Towards a Psychocivilized Society* (New York: Harper & Row, 1971).

14. J. M. R. Delgado, "Two-way Transdermal Communication with the Brain," *American Psychologist* 30 (1975): 265–73.

15. J. A. Meyer, "Crime Deterrent Transponder System," *IEEE Transactions: Aerospace and Electronic Systems* 7, no. 1 (1942): 2–22.

16. D. N. Michael. "Speculations on the Relation of the Computer to Individual Freedom—the Right to Privacy," in U.S., Congress, House, Committee on Government Operations, Special Subcommittee on Invasion of Privacy, *The Computer and the Invasion of Privacy: Hearings,* 89th Cong., 1st sess., 26–28 July 1966, pp. 184–93.

17. Schefflin and Opton, *The Mind Manipulators.*

18. J. Owen, *The Abolitionist* 7 (1981): 3–6.

19. Chavkin, *The Mind Stealers,* p. 73.

20. Department of Health and Social Services (U.K.) statistics, 1980.

21. B. F. Skinner, *Beyond Freedom and Dignity* (London: Cape, 1972).

22. Chavkin, *The Mind Stealers,* p. 79.

23. Ibid., p. 79.

24. Ibid., p. 72.

25. M. Fitzgerald and J. Sim, *British Prisons,* 2nd ed. (Oxford: Blackwell, 1981).

26. B. Coard, *How the West Indian Child is Made ESN in the British School System* (Boston: New Beacon Press, 1974); S. Tomlinson, "West Indian Children and ESN Schooling," *New Community* 6, no. 3 (1978); Camden Committee for Community Relations: evidence of the CCCR to the Rampton Committee, London, 1980.

27. S. D. Clements, *Minimal Brain Dysfunction in Children: Terminology and Identification,* U.S. Public Health Service Publication no. 1415 (Washington, D.C., 1966).

28. P. H. Wender, *Minimal Brain Dysfunction in Children* (New York: John Wiley, 1971).

29. J. S. Werry, K. Minde, A. Guzman, G. Weiss, K. Dogan, and E. Hoy, "Studies

on the Hyperactive Child. (VII) Neurological Status Compared with Neurotic and Normal Children," *American Journal of Orthopsychiatry* 42 (1972): 441–51.

30. G. Weiss, L. Hechtman, and T. Perlman, "Hyperactives as Young Adults: School, Employer, and Self-rating Scales Obtained During Ten-year Follow-up Evaluation," *American Journal of Orthopsychiatry* 48 (1978): 438–45; G. Weiss, E. Kruger, V. Danielson, and M. Elmann, "Effect of Long-term Treatment of Hyperactive Children with Methylphenidate," *Canadian Medical Association Journal* 112 (1975): 159–65.

31. R. Freeman, in *The Hyperactive Child and Stimulant Drugs*, ed. J. J. Bosco and S. S. Robin (Chicago: Univ. of Chicago Press, 1976), p. 5.

32. P. Schrag and D. Divoky, *The Myth of the Hyperactive Child and Other Means of Child Control* (New York: Pantheon, 1975).

33. G. S. Omenn, "Genetic Issues in the Syndrome of Minimal Brain Dysfunction," *Seminars in Psychiatry* 5 (1973): 5–17.

34. Bosco and Robin, *The Hyperactive Child;* also see L. A. Sroufe, "Drug Treatment of Children with Behavior Problems," in *Review of Child Development Research*, vol. 4, ed. F. D. Horowitz (Chicago: Univ. of Chicago Press, 1975); G. Weiss and L. Hechtman, "The Hyperactive Child Syndrome," *Science* 205 (1979): 1348–54.

35. J. L. Rapaport, M. S. Buchsbaum, T. P. Zahn, M. Weingartner, C. Ludlow, and E. J. Mikkelsen, "Dextroamphetamine: Cognitive and Behavioral Effects in Normal Prepubertal Boys," *Science* 199 (1978): 560–63.

36. Weiss et al., "Effect of Long-term Treatment."

37. D. P. Cantwell, "Drugs and Medical Intervention," in *Handbook of Minimal Brain Dysfunctions*, ed. H. E. Rie and E. D. Rie (New York: John Wiley, 1980), pp. 596–97.

38. J. R. Morrison and M. A. Stewart, "A Family Study of the Hyperactive Child Syndrome," *Biological Psychiatry* 3 (1971): 189–95.

39. J. R. Morrison and M. A. Stewart, "Evidence for Polygenic Inheritance in the Hyperactive Child Syndrome," *American Journal of Psychiatry* 130 (1973): 791–92.

40. M. A. Stewart, F. N. Pitts, A. G. Craig, and W. Dieruf, "The Hyperactive Child Syndrome," *American Journal of Orthopsychiatry* 36 (1966): 861–67.

41. D. P. Cantwell, "Psychiatric Illness in the Families of Hyperactive Children," *Archives of General Psychiatry* 27 (1972): 414–17.

42. J. R. Morrison and M. A. Stewart, "The Psychiatric Status of the Legal Families of Adopted Hyperactive Children," *Archives of General Psychiatry* 28 (1973): 888–91.

43. D. P. Cantwell, "Genetic Studies of Hyperactive Children: Psychiatric Illness in Biologic and Adopting Parents," in *Genetic Research in Psychiatry*, ed. R. R. Fieve, D. Rosenthal, and H. Brill (Baltimore: Johns Hopkins Univ. Press, 1975).

44. E. J. Mash and J. T. Dalby, "Behavioral Interventions for Hyperactivity," in *Hyperactivity in Children*, ed. R. L. Trites (Baltimore: University Park Press, 1979).

45. S. B. Campbell, M. Schleifer, G. Weiss, and T. Perlman, "A Two-year Follow-up of Hyperactive Preschoolers," *American Journal of Orthopsychiatry* 47 (1977): 149–62; also see S. B. Campbell, M. W. Endman, and G. Bernfeld, "A Three-Year Follow-up of Hyperactive Preschoolers into Elementary School," *Journal of Child Psychology and Psychiatry* 18 (1977): 239–49.

46. M. M. Helper. "Follow-up of Children with Minimal Brain Dysfunctions: Outcomes and Predictors," in *Handbook of Minimal Brain Dysfunctions*, ed. H. E. Rie and E. D. Rie (New York: John Wiley, 1980).

47. M. Schleifer, G. Weiss, N. Cohen, M. Elman, H. Cvejic, and E. Druger, "Hyperactivity in Preschoolers and the Effect of Methylphenidate," *American Journal of Orthopsychiatry* 45 (1975): 38–50.

48. T. McKeown, *The Role of Medicine* (Oxford: Blackwell, 1979); also see B. Inglis, *The Disease of Civilization* (London: Hodder & Stoughton, 1981); and B. Dixon, *Beyond the Magic Bullet* (London: Allen & Unwin, 1978).

49. E. S. Valenstein, *Brain Control: A Critical Examination of Brain Stimulation and Psychosurgery* (New York: John Wiley, 1974).

50. See the frank description of operations carried out by the doyen of American lobotomizers, W. Freeman, in the 1930s and 1940s, in *Lobotomy: Resort to the Knife* (New York: Van Nostrand Reinhold, 1982). For an account of psychosurgery as it affects an individual, see the coverage of the Margaret Chapman case in Britain—for instance, "Operation Heartbreak" in *Womans Own*, 15 March 1980.

51. Valenstein, *Brain Control*; S. Chorover, *From Genesis to Genocide* (Cambridge, Mass: MIT Press, 1979); also see P. R. Breggin, "The Return of Lobotomy and Psychosurgery," *Congressional Record*, (92nd Cong., 2nd sess.), 1972, pt. 5: 5567–77; E. S. Valenstein, ed., *The Psychosurgery Debate: A Model for Policy Makers in the Mental Health Area* (San Francisco: Freeman, 1980).

52. R. L. Sprague and E. K. Sleator, "Methylphenidate in Hyperkinetic Children: Differences in Dose Effects on Learning and Social Behaviour," *Science* 198 (1977): 1274–76; also see G. B. Kolata, "Childhood Hyperactivity: a New Look at Treatments and causes," *Science* 199 (1978): 515–17.

53. O. W. Sacks. *Awakenings* (London: Duckworth, 1973).

54. A. W. McCoy. *The Politics of Heroin in Southeast Asia* (New York: Harper & Row, 1973); also see Chorover, *From Genesis to Genocide.*

CHAPTER EIGHT / SCHIZOPHRENIA: THE CLASH OF DETERMINISMS

1. Department of Health and Social Services (U.K.) Statistics, 1981.

2. A. T. Scull, *Museums of Madness: The Social Organisation of Insanity in 19th Century England* (London: Allen Lane, 1979); also see B. Clarke, *Mental Disorder in Earlier Britain* (Cardiff, England, 1975). M. Foucault, *Madness and Civilization* (New York: Vintage, 1973). D. J. Rothman, *The Discovery of the Asylum: Social Order and Disorder in the New Republic* (Boston, Mass.: Little, Brown, 1971).

3. T. Szasz, *The Manufacture of Madness* (London: Routledge & Kegan Paul, 1971).

4. *Schizophrenia: Report of an International Pilot Study* (Geneva: WHO, 1973).

5. G. Bignami, "Disease models and reductionist thinking in the biomedical sciences," in *Against Biological Determinism*, ed. S. Rose (London: Allison & Busby, 1982), pp. 94–110.

6. B. Dixon, *Beyond the Magic Bullet* (London: Allen & Unwin, 1978).

7. H. L. Klawans, C. G. Goetz and S. Pertik, "Tardive Dyskinesia: Review and Update," *American Journal of Psychiatry* 137 (1980): 900–908; also see J. Ananth, "Drug-Induced Dyskinesia: A Critical Review," *International Pharmacopsychiatry* 45 (1979) 291–305.

8. For a critical review and discussion of biochemical models of schizophrenia see V. Andreoli, *La Terza via della Psichiatria* (Milan: Mondadori, 1980). In this massive literature, just one example may suffice of a recent molecular disease model: D. Hor-

robin, "A Singular Solution for Schizophrenia," *New Scientist* 28, no. 2 (1980): 642–45.

9. P. H. Venables, "Longitudinal Study of Schizophrenia," Paper 146 of Annual Meeting, British Association of Advanced Science (September 1981).

10. J. M. Neal and T. F. Oltmanns, *Schizophrenia* (New York: John Wiley, 1980), p. 202.

11. H. Harmsen and F. Lohse, *Bevölkerungsfragen* (Munich: J. F. Lehmanns, 1936).

12. Informal discussion in F. R. Moulton and P. O. Komoro, eds., *Mental Health*, Publication no. 9 (1939) American Association for the Advancement of Science, p. 145.

13. F. J. Kallmann, *The Genetics of Schizophrenia* (Locust Valley, N.Y.: J. J. Augustin, 1938), pp. 99, 131 and pp. 267–268.

14. F. J. Kallmann, "Heredity, Reproduction and Eugenic Procedure in the Field of Schizophrenia," *Eugenical News* 23 (1938): pp. 105–13.

15. F. J. Kallmann, "The Genetic Theory of Schizophrenia: An Analysis of 691 Schizophrenic Twin Index Families," *American Journal of Psychiatry* 103 (1946): 309–22.

16. F. J. Kallman, *Heredity in Health and Mental Disorder* (New York: Norton, 1953).

17. F. J. Kallmann, "Eugenic Birth Control in Schizophrenic Families," *Journal of Contraception* 3 (1938): 195–99.

18. D. Rosenthal, "The Offspring of Schizophrenic Couples," *Journal of Psychiatric Research* 4 (1966): 167–88.

19. F. J. Kallman, "The Heredo-constitutional Mechanisms of Predisposition and Resistance to Schizophrenia," *American Journal of Psychiatry* 98 (1942): 544–51.

20. J. Shields, I. I. Gottesman, and E. Slater, "Kallmann's 1946 Schizophrenic Twin Study in the Light of New Information," *Acta Psychiatrica Scandinavica* 43 (1967): 385–96.

21. E. Zerbin-Rüdin, "Schizophrenien," in *Humangenetik*, vol. 2, ed. P. E. Becker (Stuttgart: Thieme, 1967).

22. E. Slater and V. Cowie, *The Genetics of Mental Disorders* (London: Oxford Univ. Press, 1971).

23. D. Rosenthal, *Genetic Theory and Abnormal Behavior* (New York: McGraw-Hill, 1970).

24. I. I. Gottesman and J. Shields, *Schizophrenia and Genetics: A Twin Study Vantage Point* (New York: Academic Press, 1972).

25. H. M. Pollock and B. Malzberg, "Hereditary and Environmental Factors in the Causation of Manic-depressive Psychoses and Dementia Praecox," *American Journal of Psychiatry* 96 (1940): 1227–47. Also see G. Winokur, J. Morrison, J. Clancy, and R. Crowe, "The Iowa 500: II. A Blind Family History Comparison of Mania, Depression and Schizophrenia," *Archives of General Psychiatry* 27 (1972): 462–64.

26. E. Slater, *Psychotic and Neurotic Illnesses in Twins*, Medical Research Council Special Report Series no. 278 (London: Her Majesty's Stationery Office, 1953).

27. A. J. Rosanoff, L. M. Handy, I. R. Plesset, and S. Brush, "The Etiology of So-called Schizophrenic Psychoses with Special Reference to Their Occurrence in Twins," *American Journal of Psychiatry* 91 (1934): 247–86.

28. I. I. Gottesman and J. Shields, "Schizophrenia in Twins: 16 years' Consecutive Admissions to a Psychiatric Clinic," *British Journal of Psychiatry* 112 (1966): 809–18.

29. E. Kringlen, "An Epidemiological-clinical Twin Study on Schizophrenia," in *The Transmission of Schizophrenia*, eds. D. Rosenthal and S. S. Kety (Oxford: Pergamon, 1968).

30. M. G. Allen, S. Cohen, and W. Pollin, "Schizophrenia in Veteran Twins. A Diagnostic Review," *Archives of General Psychiatry* 128 (1972): 939–45.

31. M. Fischer, "Genetic and Environmental Factors in Schizophrenia: A Study of Schizophrenic Twins and Their Families," *Acta Psychiatrica Scandinavica*, Suppl. 238 (1973).

32. H. Luxenburger, "Untersuchungen an schizophrenen Zwillingen und ihren Geschwistern Zur Prüfung der Realität von Manifestationsschwankungen," *Zeitschrift für die Gesamte Neurologie und Psychiatrie* 154 (1935): 351–94.

33. E. Inouye, "Similarity and Dissimilarity of Schizophrenia in Twins," *Proceedings of the Third World Congress of Psychiatry, Montreal* (Toronto: Univ. of Toronto Press, 1961): 1: 524–30.

34. B. Harvald and M. Hauge, "Hereditary Factors Elucidated by Twin Studies," in *Genetics and the Epidemiology of Chronic Disease*, ed. J. V. Neel, M. W. Shaw, and W. J. Schull (Washington, D.C.: Department of Health, Education, and Welfare, 1965).

35. A. Hoffer and W. Pollin, "Schizophrenia in the NAS–NRC Panel of 15,909 Veteran Twin Pairs," *Archives of General Psychiatry* 23 (1970): 469–77

36. S. Snyder, *Medical World News*, 17 May 1976, p. 24.

37. P. Wender, *Medical World News*, 17 May 1976, p. 23.

38. S. S. Kety, D. Rosenthal, P. H. Wender, and F. Schulsinger, "The Types and Prevalence of Mental Illness in the Biological and Adoptive families of Adopted Schizophrenics," in *The Transmission of Schizophrenia*, ed. D. Rosenthal and S. S. Kety (Oxford: Pergamon, 1968).

39. S. S. Kety, D. Rosenthal, P. H. Wender, F. Schulsinger, and B. Jacobsen, "Mental Illness in the Biological and Adoptive Families of Adopted Individuals Who Have Become Schizophrenic," in *Genetic Research in Psychiatry*, ed. R. R. Fieve, D. Rosenthal, and H. Brill (Baltimore: Johns Hopkins Univ. Press, 1975).

40. D. Rosenthal, P. H. Wender, S. S. Kety, F. Schulsinger, J. Welner, and L. Ostergaard, "Schizophrenics' Offspring Reared in Adoptive Homes," in *The Transmission of Schizophrenia*, ed. D. Rosenthal and S. S. Kéty (Oxford: Pergamon, 1968), p. 388.

41. D. Rosenthal, P. H. Wender, S. S. Kety, J. Welner, and F. Schulsinger, "The Adopted-away Offspring of Schizophrenics," *American Journal of Psychiatry* 128 (1971): 307–11.

42. R. J. Haier, D. Rosenthal, and P. Wender, "MMPI Assessment of Psychopathology in the Adopted-away Offspring of Schizophrenics," *Archives of General Psychiatry* 35 (1978): 171–75.

43. P. H. Wender, D. Rosenthal, S. S. Kety, F. Schulsinger, and J. Welner, "Crossfostering: A Research Strategy for Clarifying the Role of Genetic and Experiential Factors in the Etiology of Schizophrenia," *Archives of General Psychiatry* 30 (1974): 121–28.

44. H. Paikin, B. Jacobsen, F. Schulsinger, K. Gottfredsen, D. Rosenthal, P. Wender, and S. S. Kety, "Characteristics of People Who Refused to Participate in a Social and Psychopathological Study," In *Genetics, environment and psychopathology*, ed. S. Mednick., F. Schulsinger, J. Higgins, and B. Bell (Amsterdam: North-Holland, 1974).

45. B. Cassou, M. Schiff, and J. Stewart, "Génétique et schizophrénie: ré-évaluation d'un consensus," *Psychiatrie de l'Enfant* 23 (1980): 87–201. See also T. Lidz and S. Blatt,

"Critique of the Danish-American Studies of the Biological and Adoptive Relatives of Adoptees Who Became Schizophrenic," *American Journal of Psychiatry* 140 (1983): 426–31.

46. P.M. Wender and D.R. Klein, "The Promise of Biological Psychiatry," *Psychology Today*, February 1981, pp. 25–41.

47. "Rampton Prisoner Victim of Bungle," *The Guardian* (London), 23 March 1981. Also see R. Littlewood and M. Lipsedge, *Aliens and Alienists: Ethnic Minorities and Psychiatry* (Harmondsworth, Middlesex, England: Penguin, 1982).

48. D. L. Rosenhan, "On Being Sane in Insane Places," *Science* 179 (1973): 250–58.

49. Foucault, *Madness and Civilization.*

50. P. Sedgwick. *Psychopolitics* (London: Pluto, 1982).

51. R. D. Laing. *The Divided Self* (London: Tavistock, 1960). Also see R. D. Laing, *The Politics of Experience and The Bird of Paradise* (Harmondsworth Middlesex, England: Penguin, 1969); R. D. Laing and A. Esterson, *Sanity, Madness and the Family* (Harmondsworth, Middlesex, England: Penguin, 1970); D. Cooper, *The Death of the Family* (Harmondsworth, Middlesex, England: Penguin, 1972); R. Boyers and R. Orrill (eds.), *R. D. Laing and Anti-Psychiatry* (Harmondsworth, Middlesex, England: Penguin, 1972).

52. A. B. Hollingshead and F. C. Redlich, *Social Class and Mental Illness* (New York: John Wiley, 1958). Also see J. K. Wing, *Reasoning About Madness* (New York: Oxford Univ. Press, 1978).

53. G. W. Brown and T. Harris. *Social Origins of Depression: Study of Psychiatric Disorder in Women* (London: Tavistock, 1978).

54. B. L. Reid, B. E. Hagan, and M. Coppleson, "Homogeneous Hetero Sapiens," *Medical Journal of Australia*, 5 May 1979, pp. 377–80.

CHAPTER NINE / SOCIOBIOLOGY: THE TOTAL SYNTHESIS

1. Among these, a favorable review in the widely read *Atlantic* magazine by Fred Hapgood, excerpts from which were used in later publisher's advertisements. Mr. Hapgood, at the time, was a writer for the Harvard University Public Relations Office.

2. E. O. Wilson, *Sociobiology: The New Synthesis* (Cambridge, Mass.: Harvard Univ. Press, 1975).

3. See "Getting Back to Nature–Our Hope for the Future," *House and Garden*, February 1976, pp. 65–66; "Why We Do What We Do: Sociobiology," *Readers Digest*, December 1977, pp. 183–84; "Sociobiology Is a New Science with New Ideas on Why We Sometimes Behave Like Cavemen," *People* magazine, November 1975, p. 7. For a very extensive bibliography of popular and scientific writing about sociobiology, see A. V. Miller, *The Genetic Imperative: Fact and Fantasy in Sociobiology* (Toronto: Pink Triangle Press, 1979).

4. The most widely reviewed and discussed are D. P. Barash, *Sociobiology and Behaviour* (Amsterdam: Elsevier, 1977); R. Dawkins, *The Selfish Gene* (Oxford, England: Oxford Univ. Press, 1976), and *The Extended Phenotype* (San Francisco: Freeman, 1981); D. Symons, *The Evolution of Human Sexuality* (Oxford, England: Oxford Univ. Press, 1979); and L. Tiger, *Optimism: The Biology of Hope* (New York: Simon & Schuster, 1978).

5. E. O. Wilson, *On Human Nature* (Cambridge, Mass.: Harvard Univ. Press, 1978).

6. See, for example, J. T. Bonner, "A New Synthesis of the Principles That Underlie All Animal Societies," *Scientific American* 233, no. 4 (October 1975); 129–30, 132; and G. E. Hutchinson, "Man Talking or Thinking," *American Naturalist* 64, no. 1 (1976): 22–27. (The *American Naturalist* does not ordinarily publish book reviews.)

7. By latest count there were fourteen; some examples are: *Biosocial Anthropology*, ed. R. Fox (London: Malaby Press, 1975); T. H. Clutton-Brock and P. Harvey, eds., *Readings in Sociobiology* (San Francisco: Freeman, 1978): and I. De Vore, *Sociobiology and the Social Sciences* (Chicago: Aldine Atherton, 1979).

8. For example: G. S. Becker, "Altruism, Egoism and Genetic Fitness: Economics and Sociobiology," *Journal of Economic Literature* 15, no. 2 (1977): 506; H. Beck, "The Ocean Hill, Brownsville and Cambodian-Kent State Crises: A Biobehavioural Approach to Human Sociobiology," *Behavioural Science* 24, no. 1 (1979): 25–36.

9. *Business Week*, 10 April 1978, pp. 100, 104.

10. E. O. Wilson, "Human Decency Is Animal," *New York Times Magazine*, 12 October 1975.

11. *Readers Digest*, "Why We Do What We Do."

12. *People* magazine, "Sociobiology Is a New Science."

13. R. Dawkins, *The Selfish Gene* (Oxford: Oxford Univ. Press, 1976).

14. J. Hirschleifer, "Economics from a Biological Viewpoint," *Journal of Law and Economics* 20, no. 1 (1977): 1–52.

15. D. T. Campbell, "Comments on the Sociobiology of Ethics and Moralizing," *Behavioral Science* 24, no. 1 (1979): 37–45.

16. Beck, "The Ocean Hill, Brownsville and Cambodian-Kent Crises."

17. O. Aldes, "A Sociobiological Analysis of the Arms Race and Soviet Military Intentions," unpublished manuscript, 1979.

18. J. D. Weinrich, "Human Sociobiology: Pair-bonding and Resource Predictability (Effects of Social Class and Race)," *Behavioral Ecology and Sociobiology* 2, no. 2 (1977): 91–118.

19. The first fully detailed attack on the epistemology of sociobiology and its use of the ethnographic record was M. Sahlins. *The Use and Abuse of Biology: An Anthropological Critique of Sociobiology* (Ann Arbor: Univ. of Michigan Press, 1976).

Shorter treatments are S. Washburn, "Animal Behaviour and Social Anthropology," *Society* 15, no. 6 (1978): 35–41; C. Geertz, "Sociosexology," *New York Review of Books*, 24 January 1980, pp. 3–4. A philosopher's explication of the reductionist errors of sociobiology is in S. Hampshire, "Illusion of Sociobiology," *New York Review of Books*, 12 October 1978, pp. 64–69.

20. For an explanation of the effect of allometric growth on tooth size, see S. J. Gould, *Ontogeny and Phylogeny* (Cambridge, Mass.: Harvard Univ. Press, 1977).

21. R. Ardrey, *The Territorial Imperative* (London: Collins, 1967), p. 5.

22. Sahlins's *Use and Abuse of Biology* distinguishes them as the "vulgar" as opposed to Wilson's "scientific" sociobiologies.

23. Wilson, *Sociobiology*, p. 120.

24. Ibid., p. 562.

25. See Chap. 1; also see M. Barker, *The New Racism* (London: Junction Books, 1981).

26. C. B. Macpherson, *The Political Theory of Possessive Individualism* (New York: Oxford Univ. Press, 1962).

27. C. Darwin, *The Origin of Species* (1859), chap. 3.

28. P. Kropotkin, *Mutual Aid* (1902), chap. 1.

29. G. Jones, *Social Darwinism and English Thought* (Hassocks, Sussex, England: Harvester Press, 1980).

30. Frederick Engels, who thought the *Origin of Species* to be a proof of the evolution of organisms, nevertheless observed: "The whole Darwinist teaching of the struggle for existence is simply a transference from society to living nature of Hobbes's doctrine of *bellum omnium contra omnes* and of the bourgeois-economic doctrine of competition together with Malthus's theory of population. When this conjurer's trick has been performed . . . the same theories are transferred back again from organic nature into history and it is now claimed that their validity as eternal *laws* of human society has been proved. The puerility of this procedure is so obvious that not a word need be said about it." Letter to P. L. Lavrov, 12–17 November 1875. (If only that were true!)

31. Quoted in R. Hofstadter, *Social Darwinism in American Thought*, revised edition (New York: George Brazillier, 1959) p. 45.

32. Max Norden in the *North American Review* (1889), as quoted in Hofstadter, *Social Darwinism in American Thought*.

33. Wilson, *Sociobiology*, pp. 572, 575.

34. Hofstadter, *Social Darwinism in American Thought*.

35. See Sociobiology Study Group (E. Allen *et al.*), "Against Sociobiology," *New York Review of Books*, 13 November 1975, pp. 33–34, for a view from the left. Precisely the same view was taken from the other shore by Paul Samuelson, "Sociobiology, a New Social Darwinism," *Newsweek*, 7 July 1975.

36. Wilson, *Sociobiology*, p. 562.

37. R. Trivers, "The Evolution of Reciprocal Altruism," *Quarterly Review of Biology* 46 (1971): 35–37.

38. Wilson, *On Human Nature*, p. 172.

39. Wilson, *Sociobiology*, p. 554.

40. Wilson, *On Human Nature*, p. 3.

41. Ibid., pp. 154–55.

42. Wilson, *Sociobiology*, pp. 564–65.

43. Ibid., p. 574.

44. For a withering attack see Sahlins, *Use and Abuse of Biology*. See also "Sociobiology, a New Biological Determinism," in Science for the People Collective, *Biology as a Social Weapon* (Minneapolis: Burgess, 1977).

45. Derek Freeman. *Margaret Mead and Samoa* (Cambridge, Mass.: Harvard Univ. Press, 1983).

46. For such an attempt, see E. L. De Bruyl and H. Sicher, *The Adaptive Chin* (Springfield, Ill.: C. C. Thomas, 1953).

47. For example, Dawkins's concept of the "meme," *The Selfish Gene*.

48. Wilson, *Sociobiology*, p. 365.

49. Wilson, *On Human Nature*, p. 81.

50. Wilson, *Sociobiology*, p. 553.

51. Wilson, *On Human Nature*, p. 109.

52. Symons, *Evolution of Human Sexuality*, p. 149.

53. R. Trivers, in *Doing What Comes Naturally*, a film produced by Hobel-Leiterman, distributed by Documents Associates, New York.

54. Wilson, *Sociobiology*, p. 562.

55. Ibid., p. 563.

56. Ibid., pp. 554–55.

57. W. Lumsden and E. O. Wilson, *Genes, Mind and Culture* (Cambridge, Mass.: Harvard Univ. Press, 1981).

58. G. Dahlberg, *Mathematical Models for Population Genetics* (New York: S. Karger, 1947).

59. Wilson, *Sociobiology*, p. 553.

60. Symons, *Evolution of Human Sexuality*, p. 145.

61. Wilson, *On Human Nature*, p. 99. Note the conflation of warfare and aggression.

62. Ibid., p. 105.

63. Ibid., p. 119.

64. Wilson, *Sociobiology*, p. 575.

65. Ibid., p. 550.

66. Wilson, *On Human Nature*, p. 172.

67. Wilson, *Sociobiology*, p. 549.

68. E. O. Wilson, "Human Decency Is Animal," *New York Times Magazine*, 12 October 1975, pp. 38–50.

69. Wilson, *Sociobiology*, p. 551.

70. See, e.g., Wilson, *Sociobiology*, p. 550, for a list of traits said to have moderate heritability.

71. See Barash, *Sociobiology and Behaviour*, chap. 3.

72. *Exploring Human Nature* (Cambridge, Mass.: Education Development Center, 1973).

73. Barash, *Sociobiology and Behavior*, p. 277.

74. Symons, *Evolution of Human Sexuality*, p. 202.

75. Ibid., p. 203.

76. Ibid., p. 204.

77. W. D. Hamilton, "The Genetical Theory of Social Behaviour," *Journal of Theoretical Biology* 7 (1964): 1–52.

78. M. Ruse, "Are There Gay Genes?" *Journal of Homosexuality* 6 (1981): 5–34.

79. Trivers, "Evolution of Reciprocal Altruism."

80. Dawkins, *The Selfish Gene*, p. 202.

81. In the original statement of the principle, now part of any textbook of population genetics, see S. Wright "Evolution in Mendelian Populations," *Genetics* 16 (1931): 97–159.

82. Ibid.

83. S. J. Gould, "Positive Allometry of Antlers in the Irish Elk, *Megaloceros giganteus,*" *Nature* 244 (1973): 375–76.

84. For experimental data and analysis see: D.S. Falconer, *Introduction to Quantitative Genetics* (New York: Ronald Press, 1960), pp. 140–49.

85. For an extensive discussion of this neutralist-selectionist controversy see: M. Kimura and T. Ohta, *Theoretical Aspects of Population Genetics* (Princeton, N.J.: Princeton Univ. Press, 1971), and R. C. Lewontin, *The Genetic Basis of Evolutionary Change* (New York: Columbia Univ. Press, 1974).

Despite Barash's attempt to dismiss the issue in one sentence and to pretend that "most opinion favors" direct selection for characters (*Sociobiology and Behaviour*, p. 53), this has been *the* leading issue in the technical and review literature of evolutionary genetics for 20 years.

86. Interview with E. O. Wilson by C. Fischler in *Le Monde*, 24 February 1980, p. 15.

1. C. J. Lumsden and E. O. Wilson, *Genes, Mind and Culture* (Cambridge, Mass.: Harvard Univ. Press, 1981).

2. C. J. Lumsden and E. O. Wilson, "Spectrum," *The Sciences* 21, no. 8 (1981).

3. M. Midgley, *Beast and Man: The Roots of Human Nature* (Hassocks, Sussex, England: Harvester Press, 1979).

4. As indeed are biological determinists when they want to escape their own traps. For examples, see the last chapter of R. Dawkins, *The Selfish Gene* (New York: Oxford Univ. Press, 1976), or E. O. Wilson *On Human Nature* (Cambridge, Mass.: Harvard Univ. Press, 1978), or D. P. Barash, *Sociobiology and Behaviour* (Amsterdam: Elsevier, 1977).

5. K. Marx. *Theses on Feuerbach* (1845), and K. Marx and F. Engels, *Selected Works* (Moscow: Progress Publishers, 1969), vol. 1.

6. R. Dawkins, *The Selfish Gene* (Oxford: Oxford Univ. Press, 1976).

7. For a collection of these epistemologies and criticisms of them, see H. Plotkin, *Evolutionary Epistemology* (New York: John Wiley, 1982).

8. R. Dawkins, *The Extended Phenotype: The Gene as the Unit of Selection* (San Francisco: Freeman, 1981).

9. J. Piaget, *Six Psychological Studies* (New York: Random House, 1967), pp. 63–64.

10. See, e.g., the collection edited by A. Koestler and J. R. Smythies, *Beyond Reductionism* (London: Hutchinson, 1969).

11. R. W. Sperry, "Mental Phenomena as Causal Determinants in Brain Function," in *Consciousness and the Brain,* ed. G. Globus, G. Maxwell, and I. Savodnik (New York: Plenum, 1976), pp. 247–56.

12. See, e.g., S. J. Gould, *Ontogeny and Phylogeny* (Cambridge, Mass.: Harvard Univ. Press, 1977).

13. Wilson, *On Human Nature.*

14. Lumsden and Wilson, *Genes, Mind and Culture.*

15. Dawkins, *The Selfish Gene.*

16. K. R. Popper and J. C. Eccles, *The Self and Its Brain* (London: Springer, 1977).

17. W. Penfield, *The Mystery of Mind* (Princeton, N.J.: Princeton Univ. Press, 1975).

18. Lumsden and Wilson, *Genes, Mind and Culture.*

19. See, for example, R. Herrnstein, *IQ and the Meritocracy* (Boston: Little, Brown, 1971).

20. Dawkins, *The Selfish Gene*, p. 21.

21. E. O. Wilson, *Sociobiology: The New Synthesis* (Cambridge, Mass.: Harvard Univ. Press, 1975), p. 561.

INDEX

Bacon, Sir Francis, 41
bacteria, 12, 275
Barash, D. P., 178, 259, 268, 310n
Bateson, Gregory, 230
beavers, 273n, 274
behavior:
   control of, 171–78
   criminal, see criminal behavior
   inheritance of, 56, 72, 251–55
   origin of, 55–57, 60
   quantification of, 51–60
   reductionist view of, 55, 60, 142, 173,
      235–36, 283
   reification of, 90–91, 248–49
Behavior Control units, 78, 177
behaviorism, 20, 78, 175, 234n, 266–67
behavior modification, 175–78
Belgium, 66, 122
Bellow, Saul, 276
Beveridge, Lord William, 79
Binet, Alfred, 15, 84–85, 89
biological determinism, 3–15
   appeal of, 10
   as "bad science," 30–36
   as "bad therapy," 188–93
   cultural determinism vs., 10, 68, 200
   defined, 6
   heritability in, see heritability
   history of, 24–28, 37–61
   inevitability in, 6, 71, 116, 135, 264
   intrinsic vs. inherited qualities in, 71
   in literature, 17–18, 24–25, 29, 56
   New Right and, 7–8, 27
   origin of, 37–61
   politics of, 7–9, 17–36, 264
   principles of, 7, 12, 68–74, 88, 135, 158,
      206, 285
   racism and, 7–8, 19, 71n, 84
   ultimate goal of, 173
   universality claimed by, 158, 244,
      245–46, 255
   universities and, 29–30
Biological Psychiatry, 184
biologism, see biological determinism
biology:
   materialist, 46–51
   modern, origin of, 45
   molecular, 57–60, 188
   of sex, 149–52

biometry, 26, 35, 51
birds, 145, 255–56, 274
blacks, 21–23, 134, 140n
   alleged inferiority of, 19, 23, 27–28, 86,
      143n
   blood groups of, 124, 125
   brain size of, 52, 143
   equality sought by, 4, 20, 21–22, 63,
      68
   infant mortality rate of, 67
   IQ testing of, 19, 117–18, 127
   mental illness and, 167, 228
   as slaves, 63, 65, 66
Blau, P., 71
blood types, 120, 122
Boring, E. G., 90, 91
bourgeois society, 238, 240–41, 245, 309n
   in eighteenth century, 25, 42, 49
   ideology of, 37–61, 74
   nature viewed by, 42–45, 72–73
   power inherited in, 72, 80, 83, 93–94
   rise of, 39–41, 63–66
   scientific ideology of, 41–42
   in seventeenth century, 5, 241
brain, 46
   of animals, 144–45
   biological determinist view of, 7,
      20–21, 23, 142–49, 198, 283
   electrode implantation in, 171, 172, 190
   evolution of, 145, 255
   functions of, 47, 51, 52–53, 144
   lateralization of, 144–49
   liaison, 283
   male vs. female, 52, 131, 139, 142–49
   mind vs., 283–84
   minimal dysfunction of, see MBD
   regions of, 52, 144, 146, 151, 153, 170,
      190, 206, 283
   size of, 52–53, 142, 144, 255
   "triune," 145
   violence and, 168–78, 190
brain damage, 53, 170, 204
Brave New World (Huxley), 93
British Journal of Psychology, 87, 105–6
British National Health Service, 204
British Psychological Society, 106
Broca, Paul, 25, 143, 144
Brown, G. W., 230
Brucke, Ernst von, 47

R. C. Lewontin, an evolutionary geneticist, is Alexander Agassiz Professor of Zoology and Biology at Harvard University, and is on the faculty of its School of Public Health. His previous books include *The Genetic Basis of Evolutionary Change* and *Human Diversity*, and he contributes frequently to *The New York Review of Books*.

Steven Rose, a neurobiologist, is chairman of the biology department and head of brain research at the Open University, in England. He is the author of *The Conscious Brain* and co-author of *Science and Society*, *The Radicalisation of Science*, and *The Political Economy of Science*.

Leon J. Kamin is a professor of psychology at Princeton University. His previous books include *The Science and Politics of IQ* and *The Intelligence Controversy*.